Computational Modelling with Single Prompts

Computational Modelling with Single Prompts provides an overview of computer models of physical phenomena and more. Each chapter is about a different model and contains a complete description, together with the algorithm and the result of the operation. There is no mathematics extended to its limits. The book provides a complete description, allowing readers to implement the algorithm themselves. The artificial intelligence (AI) elements in the book relate to the computer implementation and the programs that have been generated on the basis of the algorithms written. The results of these programs have been used as illustrations in the book. The book is mainly of use to graduate students working in computational physics, biology, chemistry, and computer science.

Key Features:

- Provides a large collection of very diverse models, the compilation of which is unique.
- Includes elements of AI and the use of large language models (LLM).
- Consists of a collection of new ideas and additional tasks to be completed independently by the readers.

Maciej Matyka, born in 1980 in Wrocław, is a computer physicist and programmer. He is a graduate of computer physics at the Faculty of Physics and Astronomy, University of Wrocław, where he is currently a research and teaching fellow at the Institute of Theoretical Physics. His professional interests include fluid flow through porous media and computer simulation methods. He independently programs algorithms he has developed and is a practitioner. He enjoys projects and working with students. He is also interested in programming and new technologies in computer science – parallel computing and machine learning. He has always had a passion for graphics and animation based on physical modeling, which he uses on computer demoscene. He is the author of numerous scientific publications in computational physics and fluid dynamics, two stand-alone books (including a book on computer simulations), and several chapters in books published in Poland and abroad.

Series in Computational Physics

Series Editors: Steven A. Gottlieb and Rubin H. Landau

Parallel Science and Engineering Applications
The Charm++ Approach
Laxmikant V. Kale, Abhinav Bhatele

Introduction to Numerical Programming
A Practical Guide for Scientists and Engineers Using Python and C/C++
Titus A. Beu

Computational Problems for Physics
With Guided Solutions Using Python
Rubin H. Landau, Manual José Páez

Introduction to Python for Science and Engineering
David J. Pine

Thermal Physics Tutorial with Python Solutions
Minjoon Kouh and Taejoon Kouh

Electrodynamics Tutorials with Python Simulations
Taejoon Kouh and Minjoon Kouh

Computational Modelling with Single Prompts
Maciej Matyka

For more information about this series, please visit: https://www.crcpress.com/Series-in-Computational-Physics/book-series/CRCSERCOMPHY

Computational Modelling with Single Prompts

Computational Modelling with Single Prompts provides an overview of computer models of physical phenomena and more. Each chapter is about a different model and contains a complete description, together with the algorithm and the result of the operation. There is no mathematics extended to its limits. The book provides a complete description, allowing readers to implement the algorithm themselves. The artificial intelligence (AI) elements in the book relate to the computer implementation and the programs that have been generated on the basis of the algorithms written. The results of these programs have been used as illustrations in the book. The book is mainly of use to graduate students working in computational physics, biology, chemistry, and computer science.

Key Features:

- Provides a large collection of very diverse models, the compilation of which is unique.
- Includes elements of AI and the use of large language models (LLM).
- Consists of a collection of new ideas and additional tasks to be completed independently by the readers.

Maciej Matyka, born in 1980 in Wrocław, is a computer physicist and programmer. He is a graduate of computer physics at the Faculty of Physics and Astronomy, University of Wrocław, where he is currently a research and teaching fellow at the Institute of Theoretical Physics. His professional interests include fluid flow through porous media and computer simulation methods. He independently programs algorithms he has developed and is a practitioner. He enjoys projects and working with students. He is also interested in programming and new technologies in computer science – parallel computing and machine learning. He has always had a passion for graphics and animation based on physical modeling, which he uses on computer demoscene. He is the author of numerous scientific publications in computational physics and fluid dynamics, two stand-alone books (including a book on computer simulations), and several chapters in books published in Poland and abroad.

Series in Computational Physics

Series Editors: Steven A. Gottlieb and Rubin H. Landau

Parallel Science and Engineering Applications
The Charm++ Approach
Laxmikant V. Kale, Abhinav Bhatele

Introduction to Numerical Programming
A Practical Guide for Scientists and Engineers Using Python and C/C++
Titus A. Beu

Computational Problems for Physics
With Guided Solutions Using Python
Rubin H. Landau, Manual José Páez

Introduction to Python for Science and Engineering
David J. Pine

Thermal Physics Tutorial with Python Solutions
Minjoon Kouh and Taejoon Kouh

Electrodynamics Tutorials with Python Simulations
Taejoon Kouh and Minjoon Kouh

Computational Modelling with Single Prompts
Maciej Matyka

For more information about this series, please visit: https://www.crcpress.com/Series-in-Computational-Physics/book-series/CRCSERCOMPHY

Computational Modelling with Single Prompts

Maciej Matyka

CRC Press
Taylor & Francis Group
Boca Raton London New York

CRC Press is an imprint of the
Taylor & Francis Group, an **informa** business

Designed cover image: by Maciej Matyka (prompting, running) with code generated in
Claude Sonnet 4 (see Chapter 29)

First edition published 2027
by CRC Press
2385 NW Executive Center Drive, Suite 320, Boca Raton FL 33431

and by CRC Press
4 Park Square, Milton Park, Abingdon, Oxon, OX14 4RN

CRC Press is an imprint of Taylor & Francis Group, LLC

ISBN: 978-1-041-22360-3 (hbk)
ISBN: 978-1-041-22810-3 (pbk)
ISBN: 978-1-003-73406-2 (ebk)

DOI: 10.1201/9781003734062

Typeset in Times
by Apex CoVantage, LLC

I dedicate this book to my ♡ family.

I dedicate this book to my ♡ family.

Contents

Introduction

Julian Tuwim, a famous Polish poet who lived from 1894 to 1953, was known for his incredible linguistic skills and his love of searching, analyzing, and collecting linguistic curiosities, phrases, and proverbs. This excellent artist is considered, to this day, one of the most outstanding Polish poets. His description of reality, even in seemingly trivial poems for children, captivates with its complexity, the sonority of words, rhymes, and structure, creating poetry in which information is conveyed on many levels. Reading Julian Tuwim's poetry is a pleasure, as it invites us to imagine what the author saw, felt, and wanted to convey. Tuwim's poems have become an integral part of Polish culture, hold a permanent place in the curricula of Polish elementary schools, and are likely ahead of their time more than we think. No Polish person would not have heard at least one poem by Tuwim.

Working with a language model to write a program with it using a single prompt is like a game in which two people (let us call them A and B) have their backs turned to each other, with A trying to explain to B what the latter should draw. In this game, we have five images: 1-original, which is the picture that person A sees; 2-description, which is the words that describe the original to the best of person A's ability – the description is transmitted to person B; 3-image in person B's head, created based on the description transmitted from person A; and 4-image on person B's paper, which is created by realizing the description and person B's idea of what the description describes. The difficulty in this game is that person B does not see the original and relies on person A's perception, who creates his description of the drawing based on his experience, sense of aesthetics, and his own assessment of the importance of the elements. Persons A and B are different, so "big" in person A's perception may be something different than "big" for person B. Mood, skill, and knowledge also matter. A precise, concise description that contains no irrelevant information helps create a picture that closely resembles the original. If we distract person B with irrelevant details, he will forget some (sometimes key) elements of the description and end up creating a weak caricature.

This description applies to this book and the concept of a single prompt. The original 1 is the non-existent, ideal solution (program) that we would like to have and aspire to. It is the original that we must describe for the language model. Description 2 is the prompt we create to describe the best possible goal for the language model, and 4 is the resulting code created by the language model based on our prompt. The difficulty in working with the language model in this methodology is manifold. First, when describing the original, we cannot see it (if we had a solution, why would we need a language model?). Second, our linguistic skills limit us on many fronts: vocabulary, the language we write in, the availability of specific terms in only one language, the length of statements, and the need to keep an eye on detail to make them specific enough to leave no doubt about what we expect. Third, the model is not a programmer and does not know whether what it has created generates the results we expect. Assuming the language model has the best of intentions, what it creates will be a better or worse caricature of our original. We aim for the best one. The precision of our prompt will determine whether the result will satisfy us. Any "hallucinations" often mentioned in this context are an expression of a kind of goodwill on the part of the language model, which tries to fill in imprecise statements and gaps in our description either with its knowledge or creativity, sometimes simply guessing the solution to an unsolvable problem we created by either not knowing what we want or not expressing ourselves precisely enough.

Therefore, a large part of the work and methodology described in this book involved repeating prompts, verifying them, and a "prompt polishing" procedure.

DOI: 10.1201/9781003734062-1

The precision of description in a compact form – the ability to include the maximum information in a small text, the ability to use language, its structure, melody, rhymes, the precise meaning of phrases and abbreviations – was the domain of the aforementioned Julian Tuwim, who, most likely, with his well-known love of science, would have excelled in writing prompts for language models.

ACKNOWLEDGMENTS

I would like to thank everyone in turn, without any special comment. It will simply be a pleasure for you, my dear ones, to be included in this book as people who, at different stages and for different reasons, helped me – sometimes with a good word, sometimes with an idea, sometimes with a tip, or just by being nearby when I needed. Some people don't realize it, but I know what their contribution was. So, thank you, simply: Klaudia Matyka, Dawid Strzelczyk, Szymon Kaczmarczyk, Piotr Nyczka, Hans Herrmann, Maria Matyka, José Soares Andrade, Humberto Carmona, Andrzej Nowiński, Sahrish Batool Naqvi, Piotr Szymczak, Tomasz Szawełło, Stanisław Żukowski, Kacper Matyka, Bartosz Matyka, Robert Matyka, Karol Matyka, Krzysztof Graczyk, Remigiusz Durka, Ziemowit Malecha, Tadeusz Tomczak, Artur Fabrycy, Zbigniew Koza, Janusz Szwabiński, Katarzyna Sznajd-Weron, Arnold Górka, Andrzej Pękalski, Bastien Chopard, Jonas Latt, Kamil Trojan, Michał Dzikowski, Tomasz Greczyło, Ewa Dębowska, Anna Bukiewicz-Szul, Sebastian Salata, Jarosław Gołembiewski, Jaromir Kopp, Marcin Wrzeszcz, Adrian Siemieniak, Koniu, Grzegorz Adamus, Adam Kolenda, Mark Ollila, Gregor Kosec, Bo Liu, Mohammad Reza Morad, and Arzhang Khalili. Thank you all.

WHAT YOU WILL FIND IN THE BOOK

When I was first preparing to give a "modelling workshop" – a regular lecture on computer modeling for physicists – I came up with the idea that I would list as many models as possible that I used, knew, researched, or had heard about or seen the effects of somewhere. My ambition was to discuss each step in detail, some longer and some shorter. Implementation was an essential element for me. When I wrote down a given algorithm in the form of code, verified that it worked, ran it, and generated some results, I fully understood it. This was the case until 2025. It has changed now.

The large language models (LLMs) that have emerged have dramatically changed my perception of code and how to work with physical models of reality. This book comes at a time when the Chinese DeepSeek model has completely changed my view of what "books 2.0," or LLM models, can do and what they can be used for. I make no secret of the fact that I am just getting to know this new tool. I am by no means an expert, but I want, together with you, to go through this new environment in a way that I have invented and see what can be achieved with it. I invite you to join me on a journey through computer reality modeling.

One modeling approach that can also be used is to create digital twins. The point of digital twins is to create an exact, digital or mathematical, copy of the system to be able to reproduce, through the evolution of the model, the behavior of every part of it, even the smallest part. This concept is interesting, with many impressive applications, but is time-consuming, effort-demanding, and resource-intensive. Such a digital twin will, of course, be able to "simulate" reality. However, the multitude of parameters and possibilities will make their influence on the system behavior slightly unclear.

In the first part, we show how to build the simplest "zero" model of a Monopoly game using the example of a Monopoly game. We will also show how interesting (valuable) information can be derived from it, even though it seems to be trivial. We then proceed to the most interesting part, which often dominates when working on various types of models and should be treated with caution. We will perform a preliminary analysis and add parameters and mechanisms to the existing game model. We will also discuss the possible changes and extensions. Based on our model, we will reveal which color on the Monopoly game board is probably the best to occupy.

Next, we discuss the stripes on the zebra. We first discuss a simple model for their formation. Whether we are talking about blobs from ink or dissolving sugar in tea, what is most interesting about these phenomena is what happens at the edge between two different substances. Therefore, we describe a simple mechanism of exchange at the edge between two substances. We will explain it in terms of a cellular automaton, an algorithm according to which many other models also work. Interestingly, the model results can represent various phenomena through visual similarity, potentially aiding in understanding their formation mechanisms, as illustrated by the title stripes on the zebra.

Then, the Game of Life comes into play. This is J. Conway's famous model, which offers fantastic possibilities for creating worlds and systems with a variety of behaviors by introducing rules for the birth, survival, and removal of cells. In this chapter, we introduce these rules, discuss how the Game of Life works, and present the result by also solving it "using pencil and paper."

In the fourth chapter, the Game of Life undergoes a metamorphosis. Instead of specifying that a cell is alive or dead, it will now have an incomplete value (from 0 to 1) of the "life" in the cell. This provides an excellent scope for parameterization, but we found that the model is stable and allows the reproducible simulation of fascinating large structures.

Subsequent chapters include more than 40 models from physics, chemistry, biology, and mathematics. These include fractals, agent simulations, sand simulations, fluid flow, a labyrinth generator based on walks without self-intersections, packed structures, bacterial clusters, a magnetism model with phase transition, and a highway traffic jam simulator (see the table of contents). This collection of models will show the richness and complexity of our world, made up entirely of different processes, the understanding of which can be preceded by modeling.

This is where the new tool comes into play. For years, I have implemented each of my simulations, models, and programs independently by hand, dedicating most of my time to setting up the working environment, designing, and identifying bugs while also learning new techniques and libraries. I suggest trying a new tool, a large language model (LLM), to support the programming part. After specifying the algorithm and documenting all specifications, the LLM can generate a fully functional computational code that implements the model, provided that the algorithm is well-written and self-explanatory. For me, this is a game-changer in computational science methodology. It is, by no means, mature. We still need to tweak, adjust, and verify the results of the LLM. However, it is so powerful and promising that I cannot ignore it.

The knowledge contained in language models is sufficient to program simple models. I have written many of the models presented in this book at some point, but many are ideas I have always wanted to implement but did not have the time to do so. For example, a maze can be composed of non-intersecting random walks. This simple algorithm is not complicated for the language model. Of course, a skilled programmer will also write it. Still, I expect to spend two to three evenings on it.

After working on this book, I have an interesting observation that I want to share with the readers. Writing an algorithm for an LLM is not simple. Among these many different models, some I knew well before writing the book, some I had only heard of, some I came up with off the top of my head based on knowledge of others, some on experience, and some by guessing. These were the ones that came out the worst, sometimes requiring several attempts at execution, even though the algorithm seemed to be written correctly – it still didn't give the expected result. The traffic jam model, for example, I first wrote down roughly the algorithm by myself and after about 10 iterations, when I arrived at a reasonably working version executed with a single prompt (the premise of this book – the prompt is supposed to be one and complete, no corrections), the language model recognized that it was dealing with a Nagel-Schreckenberg type model (I specifically write about type because I did not write it in its original form as a cellular automaton but using agent modeling). My description of the algorithm is very complicated and creates problems for the language model. Upon reviewing the original publication, I realized that the implementation of the algorithm in this model could be "prompted" in two ways. The original approach differs in principle from mine, is shorter, and works brilliantly. It is now posted as model 35 (see the prompts section).

Additionally, I have posted the working result of my previous efforts, which predated the original model, as 35a. It shows that we need to simplify our thinking in this new world and think like K. Nagel and M. Schreckenberg did in 1992! It turns out that classic works are classic for a reason. Maybe a deep understanding of the subject, model, and its workings permeates the best scientific papers, and their descriptions by leading authors in the field make their algorithms understandable by today's LLMs?

Does the LLM solve everything? Can it surprise us, replace us, and make us run out of challenges? I don't think so. First, there will still be people who want to code and program on their own because that in itself is fun (a bit like some people still transcribing books by hand). Additionally, creativity, ideas, and internal pressure will be our advantage, and this is where people will still be needed to work.

To add variety to the topics discussed, I include a few open-ended tasks in each chapter that I consider non-trivial and which will significantly enrich the analysis presented. I encourage you to experiment with LLM models and try to obtain complex and concrete answers. Some of these may still be too difficult, but some, with the help of readers, will be manageable.

I invite you to look at these new techniques through my eyes while I close this book and get back to coding!

WHY A SINGLE PROMPT? PROMPT METHODOLOGY

The concept of a single prompt is natural to me and came to mind when I first sat down to work on language models. After an unsuccessful attempt to write a poem in the style of Julian Tuwim, due to my love of simulations and computer methods in physics, my first practical idea for using language models was to write code for simulations, which I usually wrote myself, and for which I have less and less time. When I wrote my first prompt, asking it to generate the trajectory of a random walk, I was amazed that the language model wrote it down for me. This was astonishing to me, but after a while, I wanted to add to it, improve the visualization, and even calculate something on top of its results so that I could see, for example, a graph of distance over time. And then I started adding more prompts that worked on the result of the first one. "Correct this and that," "Add next to the simulation a graph that shows this and that," "It is not right, describe the axes correctly and add another graph showing the fit to the law of such and such" – these types of corrections seemed appropriate to me.

However, I do not know whether a specific neural network design causes the solution to drift in the space of possible language model responses. The more specifications I add, the less control I have over what happens in the code and in the simulation. A good solution can easily be spoiled, and models are not at all eager to maintain the good parts generated in previous prompting sessions. Therefore, I naturally began preparing separate prompts, which I supplemented with new elements while maintaining a complete description of what I am asking for. This is how I arrived at the methodology I use in this book: the first principle is that we create a single prompt. This prompt should be polished until we remove unnecessary elements, and the instruction is precise enough that the language model is clear about what we care most about. This, of course, makes working on prompts tedious, and after a while, we end up with, for example, 50 versions of the same program. This was the case with some of the models in this book, such as granular matter, driven-cavity flow, and modeling complex objects from masses and springs – these stuck in my mind the most.

I will encapsulate the methodology of my work in a few points that define the principles and the way I work and generate prompts, and which, in finishing this book, I consider important, and which I applied in my work:

Rule 1: Use a single prompt.
Provide a single prompt that comprehensively describes the requirements.

Rule 2: Use one abstraction level.
We describe the problem at the appropriate level of abstraction and are aware of it. Mixing levels of abstraction can lead to confusion and perhaps require a second prompt.

Rule 3: Human is the author.
We do not use language models to generate or revise prompts.

Rule 4: Translate prompts.
When translating a book into another language (author's note: this is also information for future translators of this book), it is imperative to translate and run the prompt in the new language (or leave the prompt in the original language). (Interestingly, recent reports suggest that Polish is the most effective language for creating the most precise prompts (Kim et al. 2025).)

Rule 5: No freedom for the model.
Do not give the model too much freedom. Be aware that running the exact prompt several times can yield different results each time, especially when the prompt is not very precise. Precision is key, and a strong, precise prompt will, more often than not, yield good results.

Rule 6: Test other models.
Ideally, every model available on the so-called market would be able to generate the right program for us based on our prompt. However, this is not the case. Trying different solutions, one can see that some models perform better at writing complex simulations, while others perform worse. Some of the models generate programs with nicer visualization, and so on. In my work, I ran my prompts in several places, testing and choosing the best one. With more complicated prompts, after dozens of attempts, one model gave the correct answer. It is then worth checking whether another will also be able to, and I get the impression that sometimes it succeeds, and that eliminating a minor loophole from the prompt allows it to run in several places.

Rule 7: No network access.
In my work, I wanted to avoid the impression that the models I describe are "copied" from some source. It is important to keep in mind that some of the models presented here are standard in my field, and "there are 1000 implementations on the Internet." In such a case, it is not difficult for the model to use the implementation in its entirety and copy the solution if it does not contain unique specifications. That is why I wrote earlier about the right level of abstraction and precision in the description, so that the model can go as low as possible and do the programmer's work for us, rather than copy the finished solution. The lack of Internet access did not make my job any easier, but a specific model provided the solution with a specific version number, which I treat as closed, having internal knowledge.

What we get by applying the aforementioned principles in the end is a concise and precise description, which, as it turns out, is enough for the language model to generate the appropriate code, and, it seems to me, should also be enough for a programmer with minimal knowledge of the field to write the program himself. This observation is important to me, so I will emphasize in this methodology that the resulting prompts are a value in themselves, and that, beyond the field of computer modeling, they can serve as a consistent description of methods, algorithms, and recipes. This is such a capsule of timeless knowledge, contained in a single prompt that serves as an instruction for a language model but can be used for entirely different purposes in the future. Who knows, perhaps a single prompt will become publishable in a scientific or technical sense, as well as become a valuable product on the open market? Does anybody want to attempt the million-dollar prompt competition?

Is there a place for prompt 2, and would it be useful? I think so. The size of a neural network is too small to support a massive creation. In my work, if my prompt was very long and contained elements at different levels of abstraction, the language model could easily have gotten lost and drifted through the solutions, paying attention to elements that are not at all important to me. My knowledge of neural networks is too limited to answer whether it is a matter of neural network size or, perhaps, the importance of the prompts themselves. Here, it is tempting to ask a second prompt that, for example, corrects the error, combines the obtained codes into a single code, or mixes different simulations together. I did not do that here, however, and I encourage readers to do their own experiments. Maybe you are the one who would write another "Computer Modeling with Second Prompt"?

The figures in the book and the cover are there to illustrate the use of the models built by the author and that the code is generated from these models.

Images illustrating models are done by author (prompting, running) with code generated in Claude Sonnet 4 (models 1, 5, 7, 10, 11, 13, 16 3D, 17, 19, 20, 23, 27, 29, 33, 35b, 37, 39, 40, 45), Claude Sonnet 4.5 (models 52, 58, 59, 60, 61, 62, 63, 64), Claude Opus 4.1 (models 15, 32, 50), Claude Opus 4.5 (models 22, 57), Grok 3 Deep Thinking (models 2, 3), Grok 3 Expert (models 4, 6, 12, 21, 34, 36), DeepSeek R1 (model 8), ChatGPT5 Deep Thinking (models 9, 14, 16 2D, 18, 24, 26, 30, 31, 38, 41, 46, 47, 53, 54), ChatGPT5.2 Extended Thinking (model 28), Gemini 2.5 Flash (models 25, 43), and Gemini 2.5 Pro (models 42, 44, 48, 49, 51, 55, 56).

1 Monopoly

Everyone has played Monopoly, Eurobusiness, or Fortune. In all these games, it is the same thing – here we move our pawns around a board of 40 fields (see picture in Figure 1.1). Movement is calculated by throwing two dice. Depending on the pawn's position, different actions can be performed. Most of the fields are various types of goods (real estate or cities, depending on the version) – we can buy them or put a hotel or other property on them. Then, the charges from the different players on that field are higher. There are special fields in the game that can take us to prison, and we can stop them (parking). In one of the fields, we have to pay a tax, and several highlighted fields are chance card selection fields with different adventures or tasks. This short description is not exhaustive, but it provides a picture of the gameplay and roughly shows that winning the game is not easy. You must adopt a strategy for buying property. Ideally, you should stand on different fields of the same color in successive passes through the board. (Only a full color allows you to place property on it.)

The game consists of a one-dimensional array of 40 elements to represent the board (see Figure 1.1). In the first approximation of our game, we will assume that we have one player, and he moves starting from the field with the number 1 by throwing a die. Movement across the board occurs cyclically. We will use a periodic boundary condition in the model, which will cause the player to be on the 1st field after passing the last field, number 40. We deliberately did not introduce additional elements from the game described earlier into the model. Our construction can now be called a "model" – a simplified creation, but one that represents a particular aspect of the system under study (perhaps a good word here would be "pretend"). The algorithm of this model consists of several elements.

Step 1. Placing the player on field number 1 (initial condition)
Step 2. Drawing a number between 0 and 6
Step 3. Moving the player
Step 4. Measuring and returning to step 2

At points 2–4, the actual game takes place. What is the measurement in step 4? This, of course, depends on what we want to measure. It is up to us, the creators and modelers, to decide what data we will read from the model and what we will track in the future. At this stage, it is essential to ask specific questions that we would like to answer. As you read this description, I suggest you close the text and think about your own; perhaps you can ask something that I have not considered. The first question that comes to mind is: how many times has a "1" fallen out? Which field is the most frequently visited? Does it matter that we started from the field "1"? Is it most difficult to stand on one of the last fields before the start? Does every number on the die fall equally often?

To answer any of these questions, it is necessary to run a simulation using the created model. The simulation should cover a large number of games and, according to the questions posed, measure and record the results of such quantities as the number of throws, the number thrown on the dice, and the field on which the player landed (these figures can be stored in a table with rows denoting successive dice throws). The simulation can be run on a computer, board, or even on a piece of paper, with the best statistics being obtained on a computer because of the large number of throws possible.

In the simplest version of the model described earlier, it is possible to predict the behavior of the model theoretically. Because we allow for single dice throws where the probability of throwing any number of dice is the same, the result of such a model and multiple passes through the board will be a uniform probability of being on each of the playing fields. Based on this simple "thought

DOI: 10.1201/9781003734062-2

FIGURE 1.1 One version of the Monopoly game (author's collection).

experiment," the above questions can be answered. "1" fell roughly 1/6 times the number of throws (1/6 being the probability of throwing one of the dice digits). For a long game, it is impossible to tell which field is the most visited because we stand on each field about the same number of times. If the game were very short (less than one pass, or at most, a few passes), then the situation would be different. Fields close to the starting field are more likely to be selected at the beginning, but this has nothing to do with statistics. With a large number of throws, the initial few moments are insignificant.

Let us now add one of the game mechanics to the model: the prison sentence. When a player stands on field 30, they are moved to the prison on field 10. The algorithm is completed by checking if the player has stood on field 30:

Step 1. Putting the player on field number 1 (initial condition).
Step 2. Drawing a number between 0 and 6.
Step 3. Moving the player.
Step 4. If the player is standing on field 30, move the player to field 10.
Step 5. Measure and return to step 2.

In this version, the player immediately leaves the prison. How does prison affect gameplay? Can we ask the same questions as those asked earlier? Of course, no one forbids us from asking questions. However, the question of how many times we throw a "1" on the dice makes no sense here, as the basic version of the model has already answered this. However, the distribution of the player's occupancy of each field becomes an interesting question. In the basic version, the filling of each

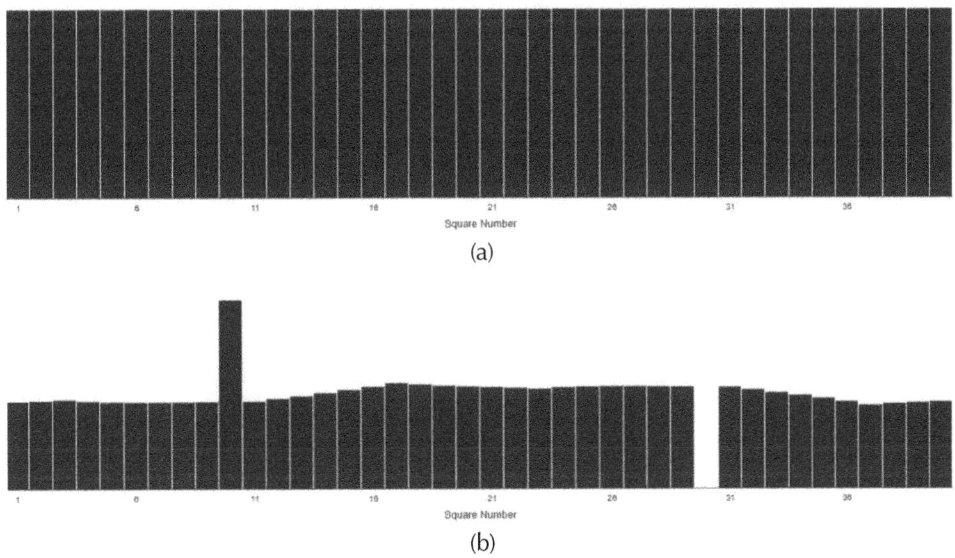

FIGURE 1.2 Histogram of the frequency of field occupancy in the Monopoly game for 103,657,000 steps. The application executes the algorithms of the game without a prison (top) and with a prison (bottom). During the operation, the successive visits to the fields in the game were continuously recorded, which is presented here in the form of a histogram for the model (program written using Prompt 1).

field was equally likely. Now, the player will also hit the prison field 1/40 times, which will move him to the prison field. This means that the prison field will have twice as many hits as the other fields. Interestingly, it can also be expected that all six fields immediately behind the prison will be visited more often because of the frequent starts from the prison. Is that all? No! Because the player will never occupy the "go to prison" field, this should also affect the fields immediately behind it.

The reasoning provided earlier is not a solution to the model. The model can be practically traced in its simplest form on a board, on a piece of paper, using a computer simulation and a specially written program, or by conducting an analysis based on probability calculus (the model in its simplest form can be described analytically).

Let us now make a simple prompt to the language model, asking for a program that models the game in question (see Prompt 1).

The result of Prompt 1 is an application that simulates the model and generates results, as illustrated in Figure 1.2. The graph shows the normalized number of player visits to a given field. The homogeneous frequency distribution in variant 1 (without the prison field) agreed perfectly with our theoretical predictions (we do not include the result as it is trivial). The heterogeneous distribution in which double the number of occurrences increases the bar on the prison field (field 10 in the figure opposite) is a consequence of the existence of the "go to prison" mechanism, and this observation also agrees with our thought experiment. Moreover, exactly six fields behind the prison had a reduced frequency, and interestingly, this frequency decreased. Meanwhile, the rest of the fields (up to the prison, where the slow increase starts) are also reduced. I leave the rest of this analysis to you, hoping that you will notice the interesting facts and clues to playing Monopoly as a result of such a simple game model with a "go to jail" mechanism only.

Task 1. Extend your analysis of the game to include additional elements familiar from the game Monopoly (e.g., going to prison after the exact numbers on both dice are drawn twice, chance cards leading to specific fields on the board, cards going back three fields, etc.). Please check the impact of each of these mechanisms and prepare a chart of the distribution of fields filled for several of them.

Task 2. Perform a series of multiplayer games according to the rules from the given algorithm and carry out an analysis of the best player strategy (e.g., simulate players who always buy properties, players who never do, intermediate cases, etc.). Conduct a complete, complex statistical analysis on a large number of games and try to conclude the correct strategy for the game.

Task 3. Develop a strategy for the game that optimizes it to its limits and allows you (still according to its rules) to gain an advantage over players who do not know it.

2 Stripes on the Zebra

I do not know how or why stripes are formed on zebras. However, similar patterns are observed in nature in many phenomena where two substances with different properties mix by exchanging molecules. Such mixing can have different mechanisms, but it is about one thing: the flow of molecules from one area to another. This exchange is the most crucial element of the model under discussion (Chopard and Droz 1998; https://en.wikipedia.org/wiki/Curve-shortening_flow).

Here, we used an algorithm called cellular automata to model the blobs. It has the feature that we perform operations on each element of our layout simultaneously. The key element is simultaneous operation. We will decide whether to light or extinguish a particular cell based on its value and those of its surroundings. To do this, we sum the values of its neighbors, add the value in the cell to it, and make a decision based on this value.

The algorithm operates on a two-dimensional board composed of equal elements. Initially, we will place the black and white values on it in a completely random manner, which is a perfect initial state for this model. In each step of the simulation, for each node, we perform a summation of the values over the neighbors and add the value of the given cell. The result is a value in the range [0, 9]. Depending on this value, the new state of the cell is selected as follows:

0 -> 0
1 -> 0
2 -> 0
3 -> 0
4 -> 1
5 -> 0
6 -> 1
7 -> 1
8 -> 1
9 -> 1

The exchange of 4 and 5 in this diagram is crucial. Such an exchange results in a transfer of value between the phases. Moreover, it turns out that, as a whole, this arrangement produces non-trivial results. Therefore, the algorithm for this model is as follows:

Step 1. Fill the array with random numbers.
Step 2. Make a copy of the array.
Step 3. For each cell of the array, calculate the sum of its neighbors and add the value from that cell (you will get the value 0–9).
Step 4. Use the rules above to select a new value for the cell and store this new value in the copy of the array.
Step 5. Once you have completed steps 3 and 4 for all the elements of the array – rewrite the copy into the array proper (this can be done element by element).
Step 6. Show the new state of the array and return to step 1.

What remains unclear here is how to treat edge cells. This is because we cannot simply count the sums after the neighbors for the edge cells in our array. After all, we have no neighbors "outside the array." There are several ways to solve this problem, and one of them is to treat the edges as cyclic (so-called periodic boundary conditions). This means that in the case of an edge cell, its neighbor

DOI: 10.1201/9781003734062-3

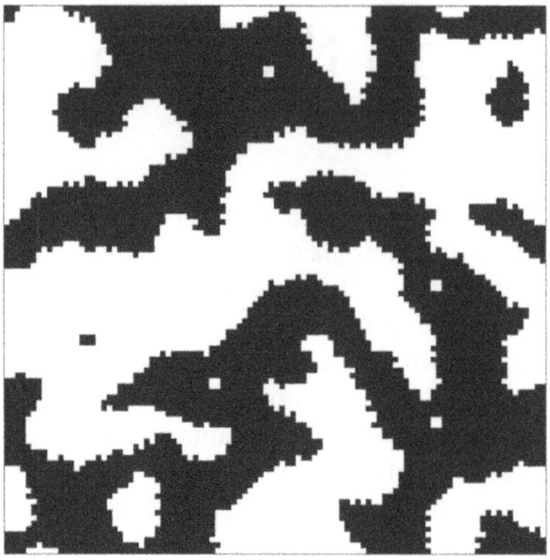

FIGURE 2.1 Effect of the stripe model made according to the proposed algorithm (program written using Prompt 2).

on the side where there are no cells will be the cells from the opposite boundary. This cyclicity will allow us to omit the problem of special treatment of boundaries. In two dimensions, the system will obtain a torus topology and no influence of edges on the solution.

The result will be a simulation that presents the evolution of the stain model. An example of the effect of such an algorithm is shown in Figure 2.1.

As can be seen in the figure, the model yields nontrivial, sizable structures with irregular shapes (initial moment). These structures, however, probably transition into a single strip spanning the whole system after some time, likely due to the boundary conditions. I encourage you to conduct experiments incorporating changes to the principles of evolution and neighbor counting in cellular automata. Perhaps you will be the one to devise a different set of rules that leads to more interesting behavior in this model.

Task 1. Please develop a three-dimensional model with similar principles and visualize it to show the formation of three-dimensional droplets.

Task 2. Perform statistical analysis of the model on multiple systems for long-simulation times. Determine the dependence of the cluster size on time and the number of clusters on time (all values averaged over multiple layouts). Please perform the analysis in 2D and 3D.

Task 3. Please perform the simulation for a given number K of steps and present the results graphically, one by one (result viewer), for random combinations of value exchange rules in the cellular automaton. Among them will probably be the one from this model, but you may randomly find another interesting model.

Task 4. Generalize the model to floating-point values and the transfer (diffusion?) of a factor at the boundary of two media, rather than their hard exchange, as in this study. Does such a model have a chance to work and ensure the formation of (soft?) patches?

Task 5. Define the model and examine its behavior on a triangular and hexagonal grid.

3 The Game of Life

The cells of living organisms divide to form successive populations, merge into larger organisms, and may die. Complex life processes are not simple to model. A prominent example of a model that has not been overcomplicated by rules but allows for the modeling of a system that resembles, in many aspects, a living system of cells that divide, merge, give birth, and die, is the Game of Life, John Conway's cellular automaton (https://en.wikipedia.org/wiki/Conway's_Game_of_Life).

Let us consider a cellular automaton model (see stain model) on a regular grid with values of 0 or 1, in which we introduce the term that a cell with a value of 1 is alive and a cell with a value of 0 is dead. The model does not differ much in operation from the simple cellular automaton of the stripes. However, we will slightly change the survival rules as follows.

1. A living cell will only survive if it has two or three neighbors.
2. A dead cell is reborn if it has three neighbors.

This will provide a complete algorithm in the following form:

Step 1. Fill the array with random numbers (or other initial state).
Step 2. Make a copy of the array.
Step 3. For each cell in the array, calculate the sum of its neighbors (do not add the value from that cell; you will get a value of 0–8).
Step 4. Use the rules above to select a new value for the cell and store this new value in the copy of the array.
Step 5. Once you have completed steps 3 and 4 for all elements of the array, rewrite the copy into the array proper (this can be done element by element).
Step 6. Show the new state of the array and return to step 1.

These principles implemented as a cellular automaton (see stain model) will realize the Game of Life model, John Conway's famous algorithm. This model has several features that distinguish it from others. These include nontrivial states and evolutions, a wealth of forms and circuits ranging from periodically changing to producing copies of themselves, solving simple algorithms, and even implementing logic gates.

To observe this model in action, consider an initial configuration in which three adjacent cells have a value of 1 and the rest of the cells around them have a value of 0. It will look like this:

0 0 0 0 0 0 0
0 0 0 0 0 0 0
0 0 1 1 1 0 0
0 0 0 0 0 0 0
0 0 0 0 0 0 0   An example of the starting state in the Game of Life.

We can now, in our head, count the individual neighbor sum values for each cell. We will now enter these as new values in the given field (assuming periodic boundary conditions).

0 0 0 0 0 0 0
|0 1 2 3 2 1 0
0 1 1 2 1 1 0

DOI: 10.1201/9781003734062-4

0 1 2 3 2 1 0
0 0 0 0 0 0 0 The sums of the neighbors are calculated around each cell in the arrangement of the initial state.

Given these values, we can now apply the algorithm (the two rules mentioned earlier). We then obtain the following configuration:

0 0 0 0 0 0 0
0 0 0 1 0 0 0
0 0 0 1 0 0 0
0 0 0 1 0 0 0
0 0 0 0 0 0 0 Status after one step of the game has been taken into effect.

Interestingly, this is a rotated initial state. After the next step of the game-of-life algorithm, we return to it, resulting in a cyclic behavior with a period of one step of the algorithm. The result is the so-called "Blinker" – a classic configuration from the Game of Life. The evolution of this setup thus looks like this (these are four consecutive steps):

0 0 0 0 0 0 0	0 0 0 0 0 0 0	0 0 0 0 0 0 0	0 0 0 0 0 0 0
0 0 0 0 0 0 0	0 0 0 1 0 0 0	0 0 0 0 0 0 0	0 0 0 1 0 0 0
0 0 1 1 1 0 0	0 0 0 1 0 0 0	0 0 1 1 1 0 0	0 0 0 1 0 0 0
0 0 0 0 0 0 0	0 0 0 1 0 0 0	0 0 0 0 0 0 0	0 0 0 1 0 0 0
0 0 0 0 0 0 0	0 0 0 0 0 0 0	0 0 0 0 0 0 0	0 0 0 0 0 0 0

The Game of Life simulates many interesting behaviors, such as sowing machines, creepers (gliders), fixed structures, blinkers, structures with different periodicity, replicating structures, object factories, and even firing guns and cartridges. We will conclude our description of this model here and leave it to the reader to further explore the richness of this model (Figure 3.1).

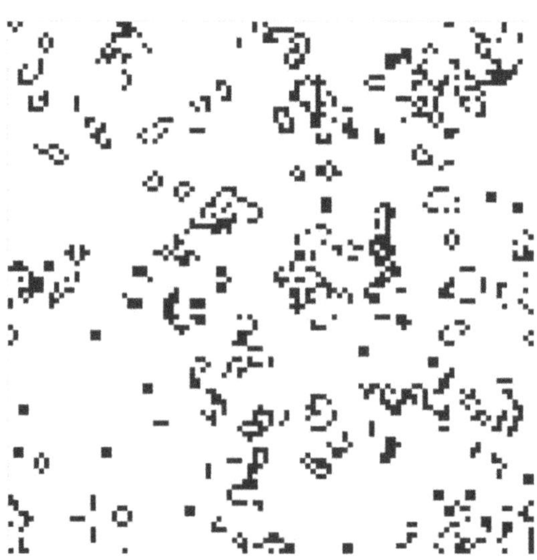

FIGURE 3.1 Evolution of the Game of Life made based on the algorithm presented. Black pixels represent living cells (program written using Prompt 3).

Task 1. Generalize the game of life to the 3D case by applying an appropriate recalculation of the birth and death rules in the model to obtain behavior similar to that of the 3D version (different types of structures, wanderers in 3D, etc.).

Task 2. Make several measurements of cell density during the simulation for different initial concentrations of living cells. Does the model always converge to the same value?

Task 3. (complex) Statistically study the system and investigate the average duration of the "life" of a structure as a function of its size. The task is not easy; it requires defining the size and making calculations with memory, that is, tagging the structures, remembering their initial size, and collecting the data at the right moment. The result should be a graph (average lifetime depending on the average size of the structure). Can you demonstrate power-law scaling in this model (see Bak et al. 1989)?

Task 4. Define the model and examine its behavior on a triangular and hexagonal grid.

4 Larger than Life

In our life game model, we used the value of a cell and its neighborhood of eight adjacent cells on a square grid. The game can be generalized to the case of a larger neighborhood, or even multiple neighborhoods. We thus obtain a model whose possibilities are basically limited by our imagination, with a wide range of parameters. These are commonly referred to as "Larger than Life" models, belonging to the class MNCA (Multiple Neighborhood Cellular Automata).

Let us consider a Game of Life model on a square grid with an extended neighborhood. Let the cell values be 0 (dead cell) and 1 (living cell). Here, we will realize an extended algorithm that, in principle, is similar to the classical game of life from the previous section. We will, however, perform the following steps:

Step 1. Fill the array with random numbers (or other initial state).
Step 2. Make a copy of the array.
Step 3 (*). For each cell, calculate the sum of the cells alive in the specified neighborhood.
Step 4 (*). Use rules that determine the new value of a cell depending on the "live" value in each neighborhood.
Step 5. Once steps 3 and 4 have been completed for all elements of the array, rewrite a copy into the array proper (this can be done element by element).
Step 6. Show the new state of the array and return to step 1.

(*) Note that only steps 3 and 4 differ from the classical algorithm in this case. To complete the description of the model, we must now address its details. We will start by defining the neighborhood.

Its choice is quite flexible – the neighborhood can be square, round, or any other shape. Here, the cell life value itself can be calculated not from a single cell but, for example, from a circle of a certain radius. Furthermore, the neighborhood can be a strip away from the cell itself, which is not taken into account at all. This makes it possible to create quite complex models in which cells interact with each other, avoid each other, or orient themselves in some sense based on the properties of their neighborhood.

To start with, let us take a quadratic environment in which the summation runs from –5 to 5 in both directions. This gives us a square area with a side length of 11 (so that the area of the whole area is 11^2 and there is a central cell in the center – see the following figure, where the central cell is in the space in the center):

```
ooooooooooo
ooooooooooo
ooooooooooo
ooooooooooo
ooooooooooo
ooooo ooooo
ooooooooooo
ooooooooooo
ooooooooooo
ooooooooooo
ooooooooooo
```

To determine the sum of the "lives" in the surroundings of the central cell, we here perform a loop over the square region defined earlier. The surroundings marked here with the letter "o" give

DOI: 10.1201/9781003734062-5

potential values of life S in the range from 0 to 120. This raises several questions, or perhaps it is better to say parameters. It is the choice of rules of thumb in the algorithm mentioned earlier (step 4), depending on the life values in the cell's surroundings, that determines how the model will behave on a macroscopic scale. For exploratory purposes, we will use the known rules for two variants. The "bugs" variant:

R=5 ("bugs" – square neighborhood)
34 <= s <= 58 -> cell survives (1->1)
34 <= s <= 45 -> cell is born (0 or 1 -> 1)

and a "stripes" variant, in which the neighborhood is circular and the conditions are similar:

R=5 ("stripes" – circular neighborhood)
34 <= s <= 58 -> cell survives (1->1)
34 < s < 45 -> cell is born (0 or 1 -> 1)

At this point, we have a complete description of the model. The algorithm, similar to the classic life game, now uses different rules depending on the value of the life boomed after considering the wider neighborhood of the cell. We will now use Prompt 4 to the LLM engine, which will generate the code that implements this simulation.

In Figure 4.1, we present the results of the Deepseek-R1 model for two different versions of this prompt, using the two sets of behaviors mentioned.

From these results, it is evident that the model is sensitive to neighborhood selection. Moreover, changing only the shape of the neighborhood caused a dramatic change in its performance and the

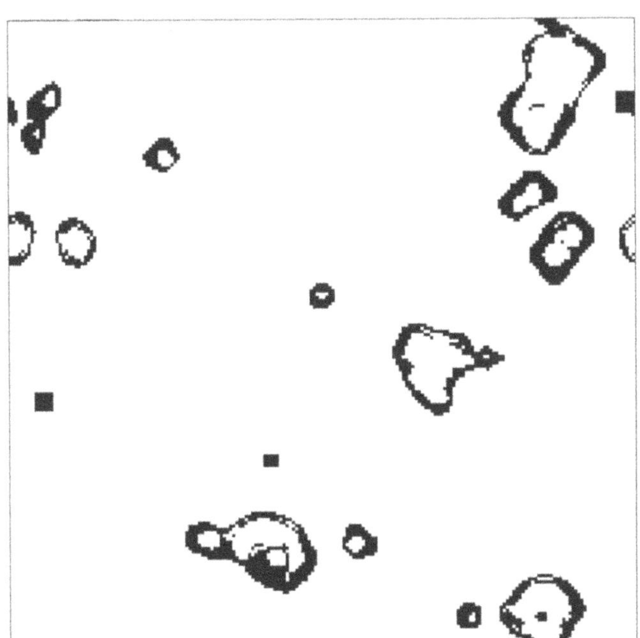

(a)

FIGURE 4.1 Two snapshots from the simulation performed with the "Larger than life" model in the "bugs" version on 400 × 400 grid, with a square (bugs, (a), initial probability of life cell p = 0.5) and circular (stripes, (b), initial p = 0.67) neighborhood. In both R = 5 was used (half-length and radius).

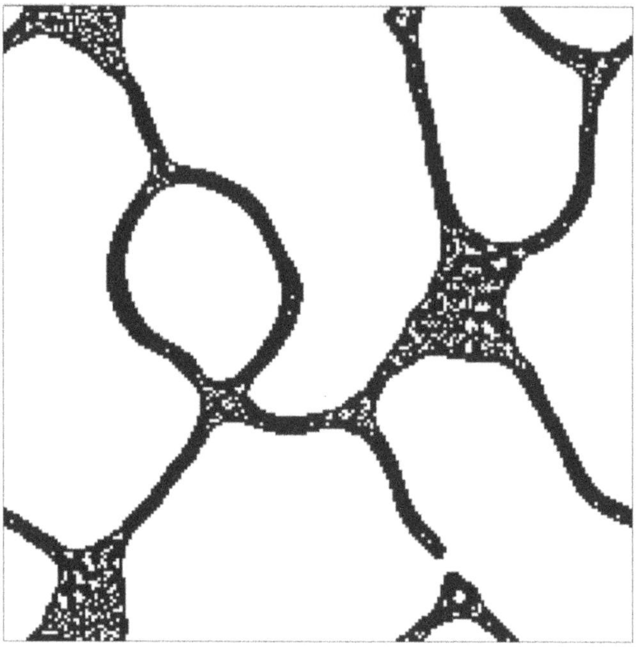

(b)

FIGURE 4.1 (Continued)

emergence of other macroscopic structures (drops vs. lines). This raises enormous possibilities, especially since the parameter space (in terms of size, shape, number of neighborhoods, and specific rules for considering them) is essentially unbounded.

Task 1. Test other neighborhoods (tire, triangle, allow the user to select neighborhood parameters and survival and birth conditions). Examine the behavior of the model in different parameter configurations.

Task 2. Generalize the model to three dimensions using cubic (spherical) neighborhoods and visualize the model using methods for visualizing three-dimensional point clouds representing living cells.

Task 3. Define the model and examine its behavior on a triangular and hexagonal grid.

5 Smooth Life

We now turn to the next class of "game of life" automaton models, in which we have more than one neighborhood (multiple neighborhood cellular automata – MNCA), here inspired by Rafler (2011). In these models, we define the surroundings of our cell using selected shapes that encompass some of our cell's surroundings, such as a circle, a square, or a tire, and other layers with a defined start and end radius. Such models are similar in algorithm and operation to the previously implemented game and are only complicated by the increased number of parameters. Whether a cell survives will now depend not only on the values in the cell, but also on the values in all defined environments. This increases the capabilities and scope of the model but also makes it increasingly difficult to control.

An additional modification we will make in this chapter will be to allow non-complete life values in cells – that is, instead of values of 0 or 1 indicating that a cell is dead or alive, from now on, there can be values of 0 to 1 in cells indicating that cells can be somewhat alive and somewhat dead. Strangely, a cell may be alive and dead, but such values can be interpreted, for example, as the age of the cell or its general condition. Such a change will also result in cells no longer changing their state abruptly but rather in gaining or losing life values. Using these values means that we can no longer apply the rules of the life game known from its discrete version, because the values will, in principle, never be total.

Let us take another look at the classic discrete version of the game of life. It turns out that the rules describing the survival or death of a given cell can be written using a simple transition matrix that defines the values in that cell for the next step.

0 0
0 0
0 0
1 1
1 0
0 0

Let us look at the figure opposite. If we plot the current state of the cell on the x-axis and the number of living neighbors on the y-axis, the matrix shows for which configuration the cell survives or is born (where the value 1 appears) or dies or remains dead (where the value zero appears). It is therefore possible to read from this matrix the same rules that we introduced in the chapter on the game of life. A living cell (first column) only survives for two or three neighbors (rows 2 and 3 from the bottom). Similarly, a dead cell (right column) is born only if it has three equally alive neighbors in its neighborhood.

Writing the transition matrix for the game of life required defining the neighborhood. Therefore, before we write a similar matrix for the continuous version, we will strictly define the neighborhood of our cell. Let us therefore introduce two surroundings of our cell. Neighborhood one will be a circular neighborhood with its center in our cell and a radius of RI (which we will call M). Surrounding two will be a rondel-shaped neighborhood adjacent to surrounding one, with a smaller radius RI and a larger radius RA.

To determine the value in a cell, we need to decide which area around the cell will be considered the neighborhood area and which area will be used to sum up the values for the cell.

Interestingly, we can be helped here by the standard game of life!

DOI: 10.1201/9781003734062-6

There, the ratio of the area of a cell to the area of the cells that surround it is exactly 1/8 (Moore's neighborhood with eight neighbors). By trial and error and experimenting with different values, a non-trivial and interesting behavior of the smoothlife model will be obtained by maintaining the same ratio of the cell area to the area of its neighborhood. Therefore, for the moment, let us assume that these quantities (in mesh units) will be RI = 4 and RA = 3*RI, which gives a ratio of 1/8 if we divide the area of the circle by the area of the tire – the surroundings marked gray in Figure 5.1.

We can now calculate the average life value of the cells in neighborhood M and neighborhood N during the simulation process of the model by summing the values and dividing by the total number of cells in both neighborhoods. Note here, of course, that the value of the cell in the neighborhood around which we calculate the averages is included in the M neighborhood, but no longer in the N neighborhood (the tire does not contain the cell in the center). We will now write down the transition matrix in the model, where, as in the classical version, we have on the x-axis the value of the life of a given cell (average of the M surroundings) and on the y-axis the value of the life in its surroundings (average of the N surroundings).

There are several variants of the continuous life game. In the figure opposite, we show the simplest of these, where the complete transition matrix is made up of two areas with stepped survival values.

Figure 5.2 shows the transition matrix for the SmoothLife model, where the x-axis represents the mean value of the surroundings M, and the y-axis represents the mean value of the surroundings N. The darker areas indicate an increase in the life value of our cell. It is therefore sufficient, after calculating both averages, to select the appropriate value S(M, N) from this matrix and apply a formula that dynamically adds or subtracts the value of life in the cell being calculated:

$$s(t+1) = s(t) + dt * (2 * S(M, N) - 1)$$

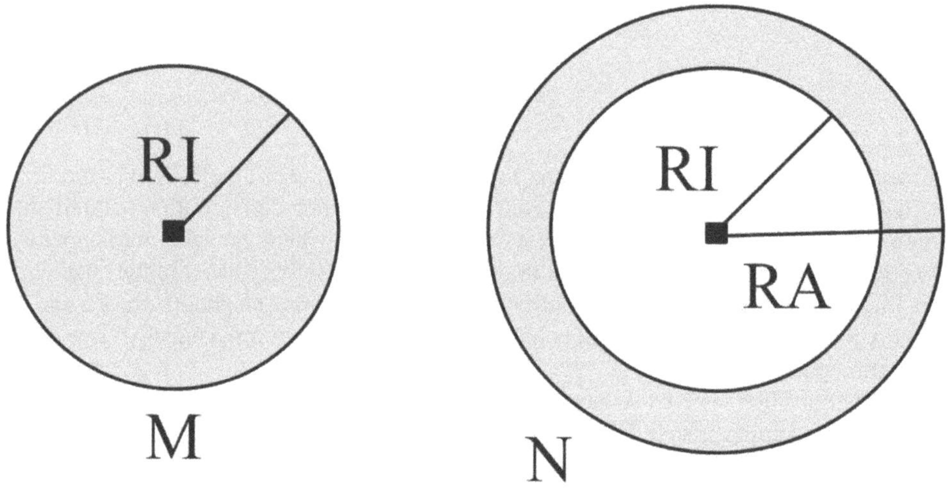

FIGURE 5.1 Neighborhood M of the cell (circular with radius RI, left), and surrounding N, in the shape of a tire, which is adjacent to neighborhood M. The cell under consideration is marked with a black square.

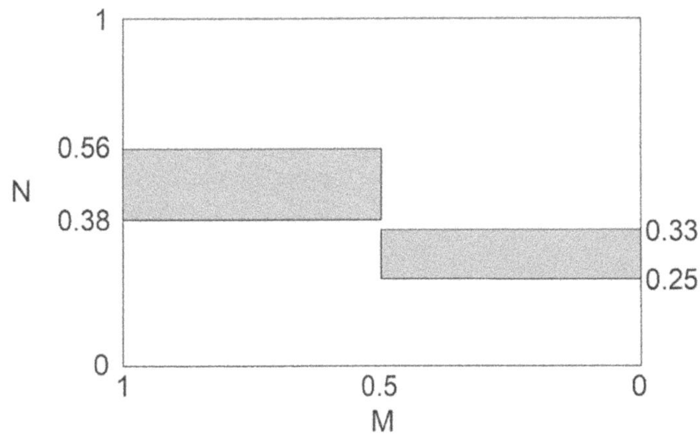

FIGURE 5.2 Transition matrix for the SmoothLife model.

In practice, this matrix can be described by two simple conditions that describe what happens to a cell depending on the values of M and N:

M<0.5 (dead) => (N > 0.25 && N<0.33)? s(t+1) = s(t) + dt: s(t+1) = s(t) – dt

M>0.5 (alive) => (N >0.38 && N<0.56)? s(t+1) = s(t) + dt: s(t+1) = s(t) – dt

The algorithm for the continuous version of the live game will therefore consist of the following steps:

Step 1. Set RI = 4 and RA = 3*RI.
Step 1a. Create a square array (e.g., 200x200) of floating point numbers and fill with random values from 0 to 1.
Step 2. For each cell i, j
2.1 Calculate M, the mean value in the circular neighborhood of radius RI (around the cell i, j),
2.2 Calculate N, the mean value in the tire-shaped neighborhood with a smaller radius equal to RI and a larger radius equal to RA (around the cell i, j).
Step 3. Using the values of M and N, apply the following conditions and determine the new value of s by writing it to a copy of the array, dt=0.01:
3.1. M<0.5 (dead) => (N > 0.25 && N<0.33)? s(t+1) = s(t) + dt: s(t+1) = s(t) – dt
3.2 M>0.5 (alive) => (N >0.38 && N<0,56)? s(t+1) = s(t) + dt: s(t+1) = s(t) – dt.
Step 4. After completing steps 2 and 3 for the entire array and creating a full copy of the array with the new values, update the primary grid with those new values.
Step 5. Visualize the grid values by color mapping them to black and white. Then, return to step 2.

The algorithm above was used to construct Prompt 5, resulting in code that generates a dynamically changing configuration of the sizable structures created by the developed algorithm (see Figure 5.3).

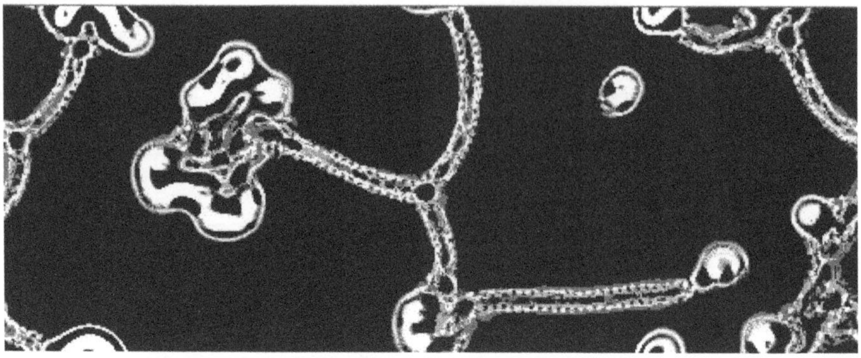

FIGURE 5.3 Example of a configuration arising in a continuous version of the SmoothLife game using the algorithm presented in this chapter (Prompt 5). Simulation done in 500 × 200 grid.

Task 1. Change the visualization to color, where you choose a color from the color map and the palette you created earlier, depending on the values of M and N in the cell.

Task 2. Add a dynamic graph of M from N, displaying the instantaneous configuration as points representing the current state of the entire system. Each point on such a graph corresponds to one cell and is located on the graph at the point M(N), where the values of M and N are the averages calculated by the algorithm for that particular cell.

Task 3. Test different arrangements of M and N neighborhoods (e.g., two tires, square-shaped tires) and try to find one that yields interesting, long-lived evolutions in the model.

Task 4. Generalize the model directly to three dimensions with visualization. Get a visualization of connected structures and the so-called glider in the SmoothLife3D model.

Task 5. Define the model and examine its behavior on a triangular and hexagonal grid.

6 Simple Waves

Already knowing how a simple cellular algorithm works and accepting that it can work on variables ranging from 0 to 1, let us try a simple simulation in which waves propagate on water. You can, of course, solve the wave equation for this purpose using, for example, its discrete form described on a differential lattice, but we will do it much more simply here. In the spirit of the cellular automata discussed, we will assume that the initial configuration is a square grid (e.g., 400 x 400), with two copies (A and B).

Let us write down the cellular automaton algorithm for the wavelet:

Step 0. Clear both grids. In array A, place 10 2x2 squares of 1200 at random locations. Copy array A to B.
Step 1. For each cell of grid A, do:
B(x, y) = (A(x−1, y) + A(x+1, y) + A(x, y+1) + A(x, y−1))/2.0 – B(x, y) //wave propagation
B(x, y) = B(x, y) * 0.9 //attenuation
Step 2. Visualization.
Step 3. Swap the arrays A and B.
Step 4. Return to step 1.

We can now ask the language model for a specific program (Prompt 6). Note that the final prompt is extended with the computation of the velocity of the wave and its visualization.

The result is a program, the results of which are shown in Figure 6.1.

Task 1. Make a 3D visualization using the deflection calculated in the 2D model as the height of the wave.
Task 2. Add obstacles to the model so that waves propagating on the water are reflected from them and try to observe the phenomenon of diffraction on the slit using this model.

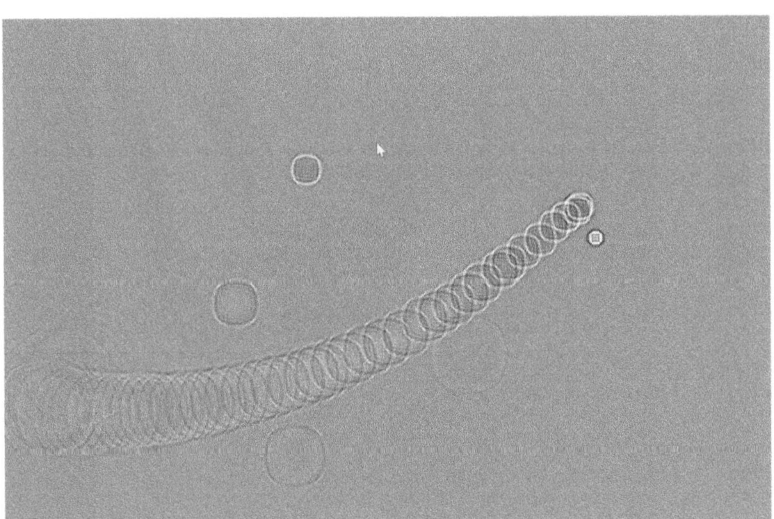

FIGURE 6.1 Waves on water modeled with a simple cellular algorithm (Prompt 6).

DOI: 10.1201/9781003734062-7

7 Lorenz Butterfly

American mathematician and meteorologist Edward Lorenz studied a system of equations to model the behavior of the atmosphere. The movements of large air masses were described by three differential equations that describe changes in quantities characterizing a system heated from below and cooled from above (analogous to what happens in our atmosphere). The original equations look like this:

$$x' = s*y - s*x$$

$$y' = -x*z + r*x - y$$

$$z' = x*y - b*z$$

However, we will not deal with atmospheric motion here, because it is not the solutions to the motion of air masses that make this system famous. It became so because it turned out that, for specific boundary conditions and parameters, this system exhibits chaotic aching. Suppose we solve this system numerically and follow its behavior by drawing the trajectory of a hypothetical particle at position (x, y, z). In that case, we will obtain a motion path resembling that of a butterfly in shape.

Let us attempt to solve this system of equations and plot the trajectory (x, y, z) using Prompt 7. As a result, we will get a program that solves the given system of equations. In Figure 7.1, you can see the solution, which is characterized by a specific transition of trajectories from the environment of one attracting solution to another.

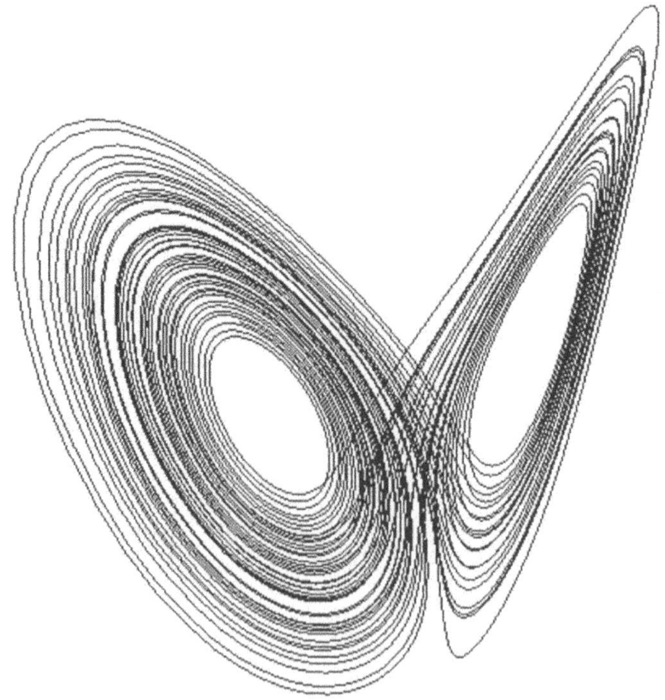

FIGURE 7.1 Butterfly – trajectory of solutions to the system of equations created by E. Lorenz.

DOI: 10.1201/9781003734062-8

Task 1. Make a different-color visualization, where the color of the drawn segment is dynamically selected from a different-color palette depending on the time of movement of the particle, its local velocity, distance from the attractors, angle of curvature of the local motion path, etc. Select the parameters in such a way as to obtain an attractive, different-colored butterfly object.

Task 2. Experiment with the Lorenz equation and try to find another stable configuration with attractors.

8 Langton Ant

How does an ant know where to go? Does it keep track of its fellows, and how is the collective movement of a large colony created (see Figure 8.1)? What role does marking the route taken by the ant play? These insects are a graceful subject for simulations. In this chapter, we will look at one of the best-known and, as usual, simple models of the movement of a single ant that can build a motorway by itself.

Imagine an ant living on a piece of checked paper. The grids are colored black and white. Depending on the color of the grid, the ant changes its orientation. If the color is white, the ant turns 90° to the left. If the color is black, it turns 90° to the right.

After a rotation, the ant makes a forward movement of one grid – this simple model, known in the literature as the Langton ant after its creator (Langton 1986).

Let us write down the complete algorithm for our ant:

Step 1. Clear the array.
Step 2. Position the agent in the center of the array and draw its direction (E, N, W, or S).
Step 3. The ant moves forward.
Step 4. Depending on the color of the field, the ant rotates:
About 90 degrees to the right if the field is white.
About 90 degrees to the left if the field is black.
Step 5. The ant re-colors the field (white <-> black).
Step 6. Visualize the array and the ant and return to step 3.

The results of the algorithm are presented in Figure 8.1. Around step 6500, we observe chaos and a somewhat disordered cluster of values on the grid with no apparent symmetry breaking. It is not easy to identify any patterns here. Meanwhile, at approximately step 10,000, an unexpected highway is generated, allowing the ant to escape from the central cluster. This is typical behavior of this model and a good test of its correct operation.

FIGURE 8.1 The trail left by an ant moving according to Langton's algorithm (step 10823) with the characteristic highway that appears at approximately step 10000 in this model (program written using Prompt 8).

DOI: 10.1201/9781003734062-9

Task 1. Generalize the model into a 3D version with the requirement that the model also results in a motorway.

Task 2. Write a model including N ants that follow one selected ant in some way (e.g., they perform their walk, but once every K steps, they make a move toward the ant they are following). The behavior of such an arrangement for a large number of ants was investigated.

Task 3. Define the model and examine its behavior on a triangular and hexagonal grid.

9 Bacterial Clusters (Eden Model)

Eden's model is a simple model of cluster growth, in which edge cells can divide, resulting in growth (Eden 1961). Let us consider the simplest version, in which we start with an embryo placed at the center of a square array. Then, at each step of the model, one edge cell is selected from the current cluster, and another is attached to it (this can be considered a division). However, we will write down the algorithm for the model with a double draw. In it, we will first draw an edge cell of the cluster; that is, one that has any empty neighbors at all, and only for this cell will we draw one of the available growth directions. We can summarize it in a few steps:

Step 1. Clear the array.
Step 2. Place an embryo in the center.
Step 3. Make a list of full cells that are adjacent to empty cells.
Step 4. Select randomly one cell from created list.
Step 5. Select randomly an empty spot from all empty spots around selected cell from previous step. Fill it.
Step 6. Update the list of full cells that are adjacent to the blank cells (make sure there are no duplicates).
Step 7. Show the cluster as a color map (black for 1 and white for 0).
Step 8. Return to step 4.

Using it, we executed Prompt 9, which created a program that generates a new type of cluster. The result is illustrated in Figure 9.1.

Task 1. Color the successive layers in the cluster using the current time to select a color from the palette. For example, the first K cells added have a color of 0, the next K cells a color with an index of 1, and so on. This will create rings in the visualization representing the cluster growth dynamics.
Task 2. Perform a simulation of growth on a triangular network. Will the model result in clusters with the same characteristics on such a network?
Task 3. Generalize the model to 3D and make a visualization of the growth of 3D clusters.

FIGURE 9.1 Eden model cluster for 65,000 cells.

DOI: 10.1201/9781003734062-10

10 Diffusion-Limited Aggregate

Imagine a stone that lies at the bottom of a slow-moving river. At the very bottom, the velocity of the river will be very low – so much so that we will neglect it. However, there will be many particles of different types in the water, and some of them (let us call them A-particles) will be able to attach themselves to this very stone. If the A-particle is in the fluid and is not lifted by the fluid, the only phenomenon that causes its movement will be diffusion – a random walk. Here, we will not consider for the moment what phenomenon causes diffusion; it can be, for example, thermal, chemical, or spontaneous. What is important instead is that such a wandering molecule of type A will, at some point, find itself close enough to our stone to attach. Then another and another. Such a model was described by Witten and Sander in 1981 in a relatively popular paper in Physical Review Letters (Witten and Sander 1981) (the paper got more than 7000 citations, which is quite a lot as for standards in physical sciences!).

However, returning to the model itself, it is extremely popular, most likely because this simple physical process, in which a molecule walks randomly in one of the available directions and attaches to the resulting cluster, leads to very complex structures with a tree-like structure, branches, a non-trivial shape, and an interesting random structure. Let us write down the algorithm for this model.

We work on a periodic table of size 400 × 400 initially cleared (value 0). In the array, write one at its center.

Step 1. Calculate the radius R of the circle described by the cluster in the center of the array.
Step 2. Draw a point (x, y) on the circle with radius R+5.
Step 3. Starting from point (x, y), perform a random walk as long as you do not hit the field adjacent to the value 1 in the array. If you hit – go to step 4. During the live walk, show the trajectory of the moving point in red.
Step 4. Write a new value of 1 in the array where the wanderer currently is.
Step 5. Show the array of values in the form of a color map – white space and a black cluster of values 1.
Step 6. If the cluster has reached the edge of the array, finish; if not, return to step 1.

After writing the algorithm in the form of a program using a slightly modified prompt, in which we added optimization in the form of tracking more wanderers at once (Prompt 10), in addition to the above algorithm, we obtained the results shown in Figure 10.1.

Task 1. Generalize the model to 3D and perform 3D tree growth with visualization.
Task 2. Add the attraction of wanderers to the center of mass of the cluster so that each random movement has an increased probability of movement toward the center of mass of the entire cluster. Create a visualization of the structures that have formed.
Task 3. Define the model and examine its behavior on a triangular and hexagonal grid.

DOI: 10.1201/9781003734062-11

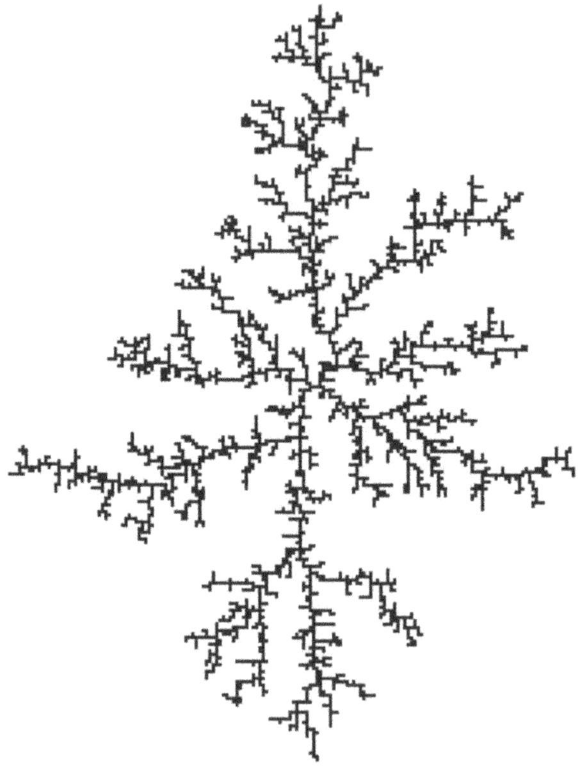

FIGURE 10.1 Diffusion-limited aggregate cluster done using introduced algorithm with program generated using Prompt 10.

11 Mandelbrot Set

Benoît Mandelbrot once told a story about how he dealt with iterations of complex functions. They did not seem interesting at all at the time, but when he drew the convergence of these iterations in the form of color maps, he was stunned because he saw fractals of the set, which today we call the Mandelbrot set.

These amazing figures remain largely unexplored to this day. In fact, I am not sure if there is anyone who can truly explain their origins. To generate them, one must study the convergence of a function that operates on complex variables. This function is described in the quadratic area and has the form:

z = z^2 + c,

where z is a complex variable, and it is this variable that is calculated in successive iterations, c is a constant (also composite, of course). Subsequent values of z are calculated in practice for successive pixels of the drawing, with zero as the initial condition for z and the coordinates of a point in the composite space for c. The choice of the limits of this space (minimum and maximum components of x and y) determines what image (what fragment of the fractal) we get. The clay limits of this space (for the characteristic Mandelbrot beetle of the title) are –2, 0.5, and –1.1, 1.1.

The runaway algorithm will check whether successive iterations of the aforementioned function escape (or not) to infinity, and based on the properties of this escape, it will color the explored space. So let us write down the complete algorithm:

We work on a 600 × 500 rectangular image. For each pixel of the image, we choose a color based on the following algorithm:

Step 1. Map the position (px, py)->(x, y) of a point in the image to variables in the range:
px->x, where x belongs to (–2, 0.5)
py->y, where y belongs to (–1.1, 1.1)
Step 2. z=(0, 0)
Step 3. c=(x, y)
Step 4. as long as z.x^2+z.y^2 <= 4 and iteration < MAXITER
{
z = z^2 + c //remember to do this, considering the complex domain
Iteration++
}
Step 5. Color the pixel with the RGB color corresponding to the iteration number (here, you can come up with some interesting color mapping).

While executing the above algorithm, draw the result on the screen in real time.

Using Prompt 11, we can generate a program that produces the Mandelbrot set, as shown in Figure 11.1.

Task 1. Modify the algorithm so that it generates fractals from the Julia set, in which the iterated function variable c becomes a constant (as opposed to the Mandelbrot set, in which c depended on the position on the screen).

Task 2. Use splines to interpolate the window bounds of the generated fractals and vary them continuously so that you get the effect of animation of the fractal figure (inlet, sliding, etc.). Using this technique, generate an animation file with smooth transitions between multiple locations within the fractal.

DOI: 10.1201/9781003734062-12

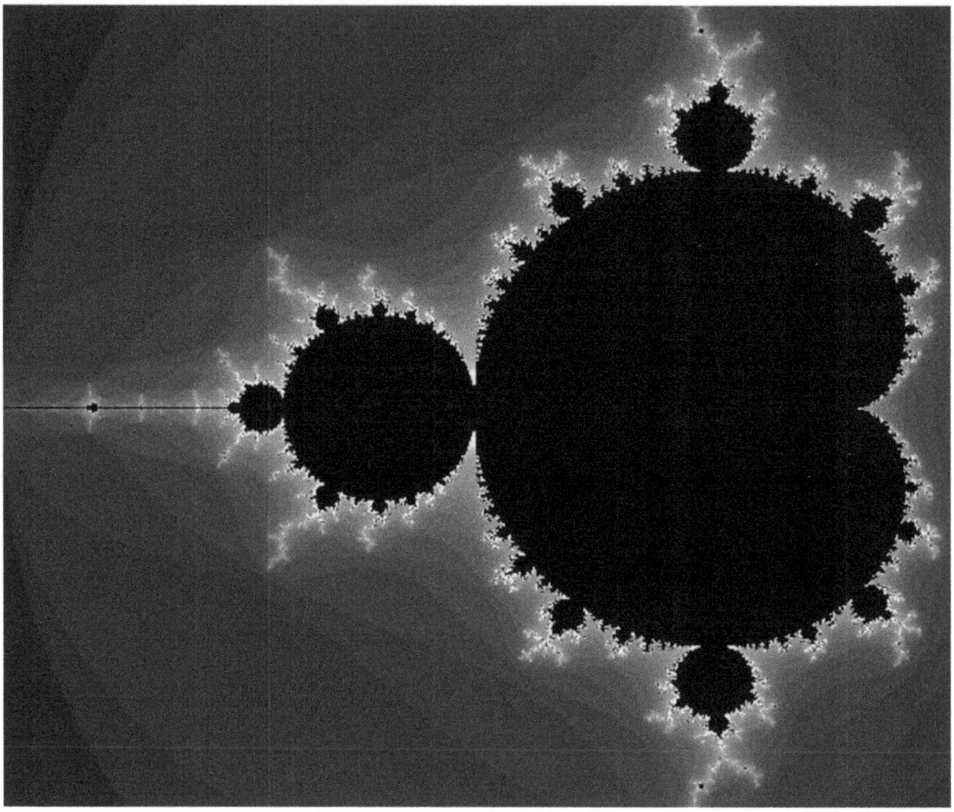

FIGURE 11.1 Mandelbrot set generated according to the given escape algorithm.

12 The Three-Body Problem

The three-body problem has fascinated people for decades. The motion of three bodies with the same masses that interact gravitationally is analytically unsolvable. Why can't we solve such a simple system? Despite the relative simplicity of the problem itself, motion in such a system can be chaotic, meaning it is highly sensitive to initial conditions. Even a slight change in the initial arrangement of bodies will lead to a completely different configuration over time.

We will model this system by directly solving the equations of motion for each of the three bodies. This system looks as follows (we assume mass = 1):

d^2r1/dt = F12+F13

d^2r2/dt = F21+F23

d^2r3/dt = F31+F33

where r and F are the vectors of velocity and force, respectively, for each of the three bodies. We define the force Fab exerted by body b on body a directly from the law of gravity as follows:

Fab = G * ma * mb * rab/rab^3,

where G is the gravitational constant (for our purposes, we can assume G = 1), the masses ma and mb are equal to each other (the masses can also be 1 for simplicity), and the vector rab is the vector from point a to point b.

These equations can be solved using one of the available numerical methods, such as the Verlet method, which allows us to count the successive positions of each body in the next step n+1 using the current position in step n and the position from the previous step n–1 according to the formula:

r(n+1) = 2*r(t) – r(t–1) + dt*dt * F(t)/m,

where m is the mass of the body whose new position we are determining, and F(t) is the sum of the forces acting on that body. Applying the above equation to all three bodies in turn, and taking care to appropriately rewrite the new position values in successive steps (a peculiarity of the Verlet method is that one must remember the positions from the previous time step as well), we can track the trajectories of their motion.

One unknown remains: how to determine the previous step in the first-time step of the Verlet method. Here, a simple integration of the Euler linear method comes to the rescue, allowing us to perform the first step (less accurately, but well enough to ensure the start of the Verlet method).

Let us gather this information into the form of a complete algorithm for the three-body problem:

Step 1. Set the positions of three bodies of the same mass at equal distances (at the vertices of an equilateral triangle).

Step 2. Perform in-loop integration of the equations of motion of these bodies:

d^2r1/dt = F12+F13

d^2r2/dt = F21+F23

d^2r3/dt = F31+F33,

DOI: 10.1201/9781003734062-13

where, in general, the Fab forces exerted by body b on body a are Fab = rab/rab^3, and the vector rab = rb-ra.

Use Verlet's method for integration:

r(n+1) = 2*r(t) – r(t–1) + dt*dt * F(t)/m, where F(t) is the sum of the forces acting on that body (from the above equations). For the first step, use Euler's method:

r(1) = r(0) + dt*dt*F(0)/m

Step 3. Visualize masses.

To test this algorithm, we will perform a simulation for the stable-eight configuration proposed in the paper (Chenciner and Montgomery 2000) – see prompt 12, here performed with DeepseekR1). Setting the points and their initial positions will make their motion regular and stagger the trajectory, as shown in Figure 12.1. However, changing the initial position by even a small amount will result in chaotic and unpredictable behavior, which can be easily observed by modifying the initial values.

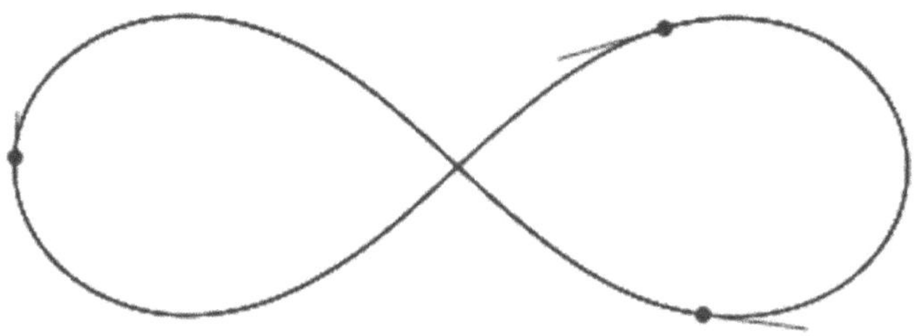

FIGURE 12.1 Loop 8 trajectory circled cyclically by three moving bodies interacting by gravity with initial conditions taken from the work of Chenciner and Montgomery (2000).

Task 1. Perform a 3D simulation with visualization and attempt to identify both stable and chaotic initial systems.

Task 2. Run a series of simulations in which one of the three masses initially located in positions for the "eight" as in the sample prompt is slightly deviated by "d" from the equilibrium position in the chosen direction. Check, starting with a minimal deviation (e.g., d = 10^–7, then d = 10^–6), how the average distance of the masses from each other changes as a function of d. Does the system already, for the slightest deviation, cease to be stable?

13 Random Walk

A random walk is a movement in which a molecule moves in each step in a random direction. This is a straightforward model that can be used to track both single molecules and groups of molecules. Such a group of molecules can represent a specific larger volume that realizes the macroscopic phenomenon of diffusion. The walk itself is a consequence of the movement of a single molecule, the random movements of which can be caused by temperature, chemical phenomena, or simply the intrinsic movements of the molecule. Without going into the genesis of motion per se, let us consider the following model of a random walk in a three-dimensional version:

Step 1. Set the position of the particle to 0, 0, 0.
Step 2. Make a random step in 3D space by choosing an angle randomly from a continuous distribution.
Step 3. Store and remember the position of the particle.
Step 4. Draw the trajectory.
Step 5. Return to step 2.

The program that implements the above algorithm can be generated, for example, with Prompt 13. It generates a visualization of a spline interpolating a random walk trajectory, as shown in Figure 13.1.

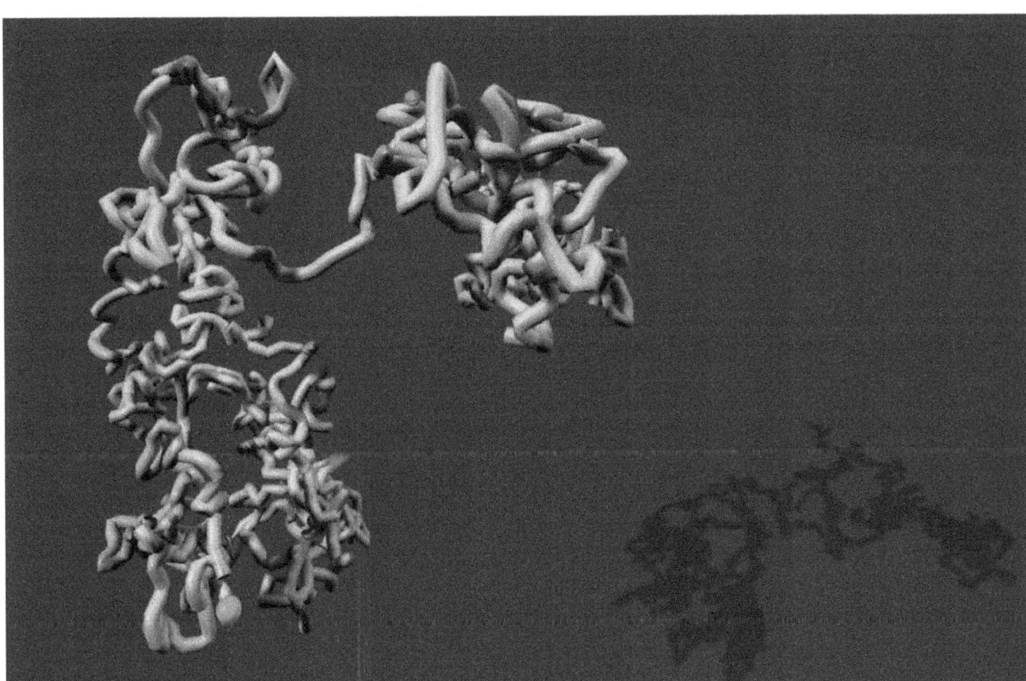

FIGURE 13.1 Trajectory in the 3D random walk simulation.

DOI: 10.1201/9781003734062-14

Task1. Take 100 independent walks of 10,000 steps each (starting from the same point) and see what the distribution of final distances from the starting point is. Does this distribution remind you of something? What function can you fit to it?

Task 2. Perform a walk on a three-dimensional grid, taking into account the condition of no self-intersections; that is, a particle cannot pass through an already visited node of the grid. Perform a walk for a large number of steps and fit a spline-type glued curve to the trajectory so formed, which approximates the subsequent positions (the curve does not have to pass through points exactly, so it should be smooth). Visualize the curve created in this way, taking into account lighting (shadows cast by the curve on itself) and global illumination (use the ambient occlusion technique).

14 Galton Board

If we drop a ball that falls on a regular set of obstacles arranged in a structure, the ball, as it falls from obstacle to obstacle, will make random movements once to the left and once to the right during free fall. The trajectory of such a ball will follow that of a one-dimensional random walk. If we throw a significant number of such balls, each slightly deviated from its initial position, each of them will follow its own trajectory and finish in various, dispersed end positions. If we count all the balls that fell and look at where, on average, there were the most of them, it is not hard to guess that the most will be in the center, and the least on the edges. A normal distribution well describes the distribution of such distances.

An interesting physical realization of such a random walk is the Galton Board (see Figure 14.1).

It is a toy that can be bought in a store, or built yourself, and consists of a board and nails (or other objects from which the ball can bounce) knocked into it in such a way that there is one nail

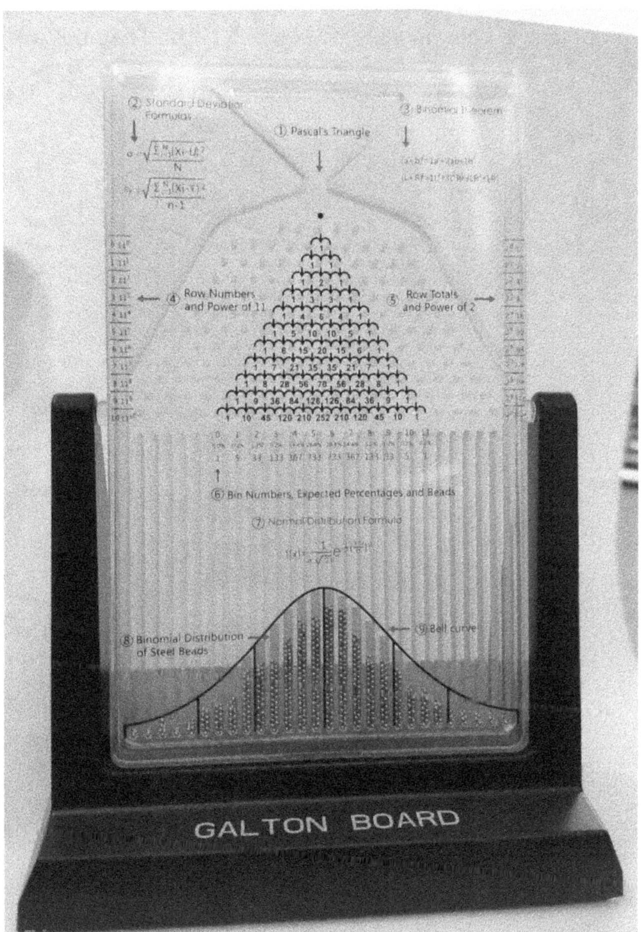

FIGURE 14.1 Galton board toy (private collection from author); notice normal distribution of balls on the bottom.

DOI: 10.1201/9781003734062-15

at the very top, two, three, etc., below in successive rows. Subsequent rows are offset by half the distance in such a way that a triangle (similar to Pascal's mathematical triangle) of successive nails is formed.

Above such a structure, we release successive balls one by one with some delay, with a slight random horizontal shift (no more than 1/10 of the radius of the nail), which physically (simulation) bounce off these nails. Such balls should bounce lightly (damping) so that they do not fall out too much, and not so much so that subsequent bounces carry them between the nails. If we count the histogram of the outgoing balls from such a system, we should see a normal distribution.

Let us write down algorithmically the implementation of the Galton Board simulation that directly models the board, balls, and the whole process of falling. For that, we need to describe the details of the board's structure as well as the simulation process (integration of ball motion equations, visualization details, and statistical measurements). We will summarize the whole simulation in the following algorithm:

Step 1. Distribute circular obstacles of radius Robs at a distance of 4Robs in a triangular mesh, so that the first row has one obstacle at the very center, the second row has two obstacles, the third row has three, and so on (total 14 rows visible).

Step 2. Place a ball (radius R) above the first row, randomly shifted +/– 0.1R in x direction.

Step 3. Simulate motion of the ball in a gravitational field. Ball collides with obstacles.

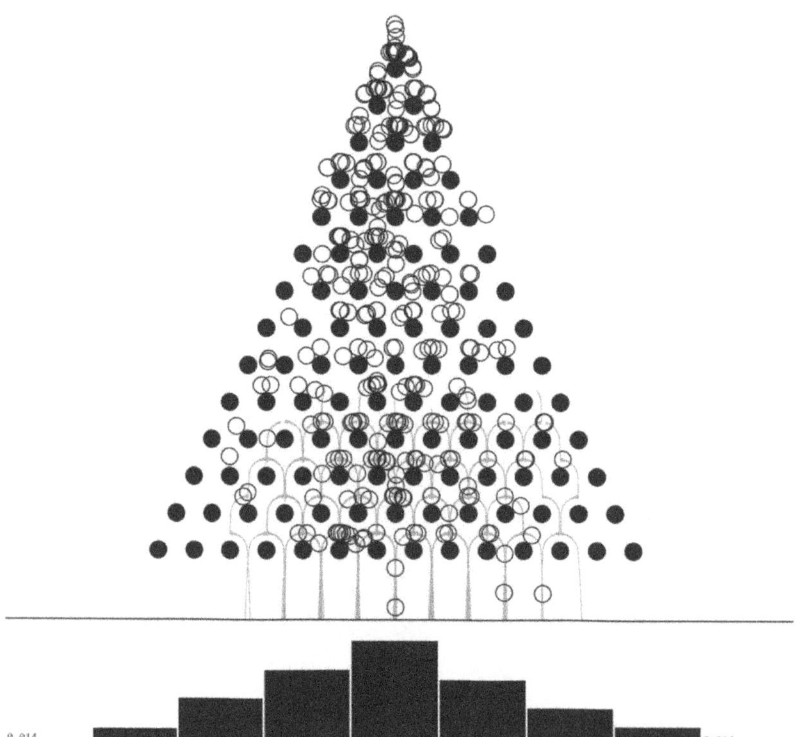

FIGURE 14.2 Galton's board model in the form of a screenshot of the model's instantaneous state and a histogram of the horizontal position of the outgoing particles (bottom graph). Falling particles are marked in black, while obstacles are marked in gray. Selected trajectories of the falling particles are also highlighted in the visualization. Parameters: damping = 1, Robs = 10, R = 9.

Step 3a. Collisions are realized exactly (ball-obstacle are two circular objects and the exact response to collision may be applied here). Use damping parameter to dissipate kinetic energy during collision. There is no ball-ball collision if many balls are released at once.

Step 3b. Count the number of balls falling on the floor of the system. Use histogram with the number of buckets equal to the number of obstacles in the last row+1.

Step 3c. If ball falls on the floor: remove it from the model and add go to step 2.

Step 3d. Visualize the obstacles, the falling ball, its trajectory, and the histogram in simple black and white form. Store the trajectories of all the balls and display them all.

Step 3e. If the ball fell on the bottom edge, remove it from the model and return to step 2. Continue the simulation.

Figure 14.2 shows the result of the Galton Board simulation conducted using the program generated with Prompt 14.

Task 1. Fit a Gaussian curve to the resulting histogram and assess the quality of the fit based on the physical parameters of the Galton Board model (e.g., damping coefficient, gravity, radius of falling balls).

Task 2. Design a Galton Board in 3D, in which balls fall on a pyramid formed by obstacles. Examine the distribution of the falling balls and see if it matches the 2D version.

15 Ising Model

Magnets have their magic. Who did not play with them as a child? However, where does the fact that they attract each other come from? Well, in a nutshell and simplification, electrons, which, among other things, make up the matter around us, have a property called spin. The spin of an electron can have one of two values; that is, be positive or negative. If one were to mix electrons with different spins, the material consisting of them would not exhibit any magnetic properties, because electrons with different spins would cancel out their interactions with each other. Such a material is called diamagnetic. However, it is possible to cause certain compact groups of electrons in a ferromagnet to align their spins in the same direction – the material will then become magnetic. This can be induced by applying a strong external magnetic field or by cooling the ferromagnetic material below the Curie temperature.

One of the best-known models of ferromagnetics is the Ising model. It consists of a regular lattice, which is filled with spins that are either +1 or −1. The basis for this model's operation is the Hamiltonian, which enables the determination of the system's energy. In the Ising model for a material without an external magnetic field, the Hamiltonian is defined as follows:

H = −J \sum_{i, j} s_i s_j

where the sum runs over all neighboring spins (only nearest neighbors are considered), the constant J should be positive for the ferromagnetic case.

To track the evolution of the system, we will employ the Monte Carlo method and the Metropolis algorithm, in which we will accept random changes in the spin configuration with a certain probability, depending on the energy difference in the system caused by this change and the temperature. In our model, we will work with random, frozen, boundary conditions (see Dasgupta et al. 1995).

So let us write down the Metropolis algorithm for simulating the Ising model of a ferromagnet:
We are working in a 60x40 array with periodic boundary conditions.

Step 1. Fill the array with +1 values.
Step 2. Randomly select one of the cells from the array.
Step 3. Invert the sign of the selected cell and count the induced energy change "DE" (directly from the Hamiltonian H = −J \sum_{i, j} s_i s_j), where s_i is the selected cell and j runs over its nearest four neighbors. To do this, you need to count the energy before and after changing the value in the cell.
Step 4. If DE < 0 -> accept the change. If DE > 0 -> accept the change with probability exp(-BETA * DE). – Repeat steps 2–4 1000 times before visualizing the array and counting the magnetization.
Step 5. Visualize the array as a black-and-white board (five times larger for better visualization).
Step 6. Go back to step 2. Assume J=1, BETA=1/T, where T is the temperature.

In the case of the Ising model, in addition to performing basic calculations and visualizations using the above algorithm, we are also tempted to construct a mechanism to study the phase transition. For this purpose, we asked the language model to create a program (see prompt 15), which additionally adds the ability to monitor the absolute value of the average magnetization in the system. We calculate this magnetization by summing the values of spins in the entire system. To ensure accurate calculation, we determine the magnetization by averaging the last 100 steps of the Monte Carlo simulation. In this way, we eliminate noise from the results. The full prompt sent to the Claude

DOI: 10.1201/9781003734062-16

Sonnet 3.7 (Perplexity) model generated the program, the results of which are presented in Figures 15.1 and 15.2.

M(T) =(1-sinh(2/T)^(−4))^(1/8)) (see Benettin et al. 1973).

Task 1. Conduct an automated analysis of the phase transition in the presented Ising model and the dependence of the M(T) curve on the system size. What minimum size of the system must be to observe the phase transition?

Task 2. Generalize the program code to three dimensions and study the phase transition in the three-dimensional model.

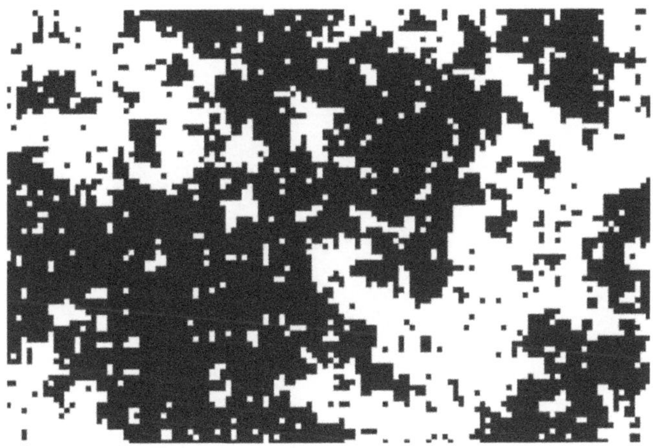

FIGURE 15.1 Spin configurations in the Ising model for temperature T = 2.3 (configuration after thermalization of the system).

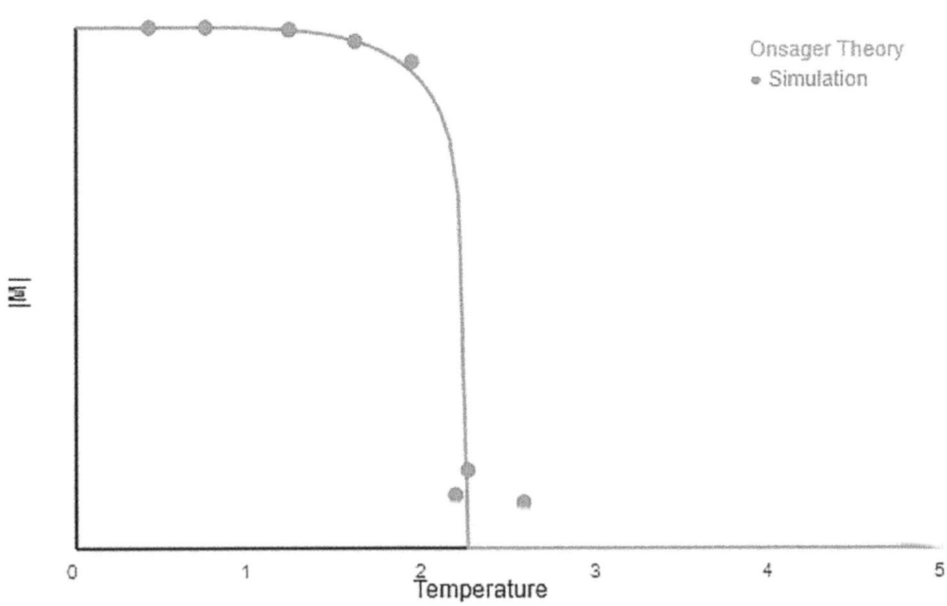

FIGURE 15.2 Measurements taken according to the presented algorithm (points) and the theoretical solution (dashed line) according to Lars Onsager's equation.

Task 3. Develop an engaging visualization for a two-dimensional model where the color of the spins depends on their history, for example, the color number corresponds to the number of spins "up" in the last 256 steps. Choose an interesting palette and an interesting function that maps the history of spins to color.

Task 4. Define the model and examine its behavior on a triangular and hexagonal grid.

16 Grains in a Silo

Let us consider the problem of packing grains in a silo. We will ask the question of how tightly and in what configuration a certain number of grains can be packed in a box. For this, we will use a simple model of growing grains, disregarding the accuracy of the physical packing process itself. Let us consider, for simplicity, a rectangular silo (e.g., 600 × 400 in dimensionless units) and round grains (let us call them disks). With this simplification, we no longer need to consider their orientation; instead, we only need to consider their position and radius. We will start modeling by placing a certain number of grains in a box and letting them slowly "grow." The growth must be very slow, and ideally, all the grains should grow together in a synchronized manner. During such growth, a situation may arise where some of the grains overlap. We then spread the overlapping grains – each in its own direction. Such a procedure is a simple realization of the collision of two objects. If, during the process of sliding apart, the disk encounters a neighbor behind it (and another collision occurs), this one will also start to move along with it. Such a procedure will enable the system to respond "globally" to the resulting collisions and relax into a densely packed state. Of course, the procedure has a limit – the disks may encounter obstacles, which, when attached to them and moved further, will encounter another and another until they reach a wall that prevents them from moving further. Then there will be a shift along the wall, but there, too, a limit can be encountered until the moment when relaxation is no longer possible. In such a situation, we will call it obtaining maximum packing and terminate the action.

Let us now write down the algorithm for the model of dense packing of round grains (rigid disks) described earlier:

Step 1. Draw N = 25 disks with radius R = W/600.0 (W – width of the canvas).
Step 2. In a loop for each point: step 2.1 The disk increases its diameter by EPS=0.01*W.
Step 2.2 Check if a disc overlaps any other:
If not, go to step 3.
If yes,
(a) move the disks apart in a loop (with a step EPS2 = 0.001*W) until there is no collision with another disk.
- if it happens – continue to move apart, together with the other disk (as if they were glued together), and then if there is a further collision – with the next one and the next one (to ensure that no pair ever overlaps).

(b) Use rigid boundary conditions at the edges (disks do not overlap at the edge when moving or growing).

Step 3. Visualize the disks in black and white (as circles) during the above procedure.
Step 4. Terminate the simulation if you get a configuration where moving is no longer possible (checking more bodies and trying to move them together leads to a situation where the disk would have to move in two directions) – then we have a situation of dense packing; return then to the last configuration where no disks overlapped and terminate the simulation.

Figure 16.1 displays the results of the above algorithm, as performed with Prompt 16 (notice that only single sentence is necessary in front of the prompt to be performed in 3D). The figure shows the characteristic hexagonal structure of densely packed grains.

DOI: 10.1201/9781003734062-17

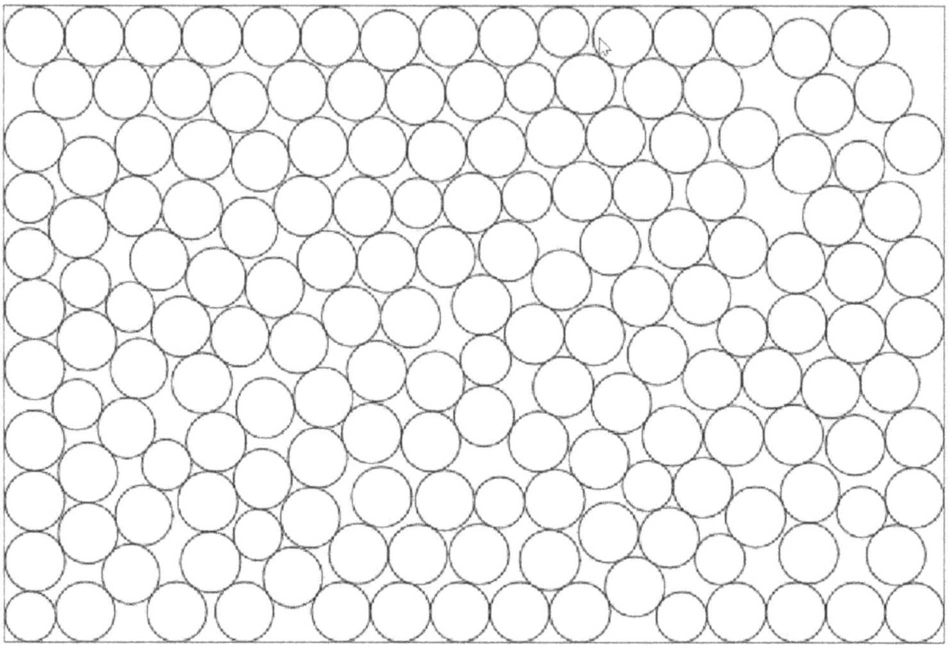

(a)

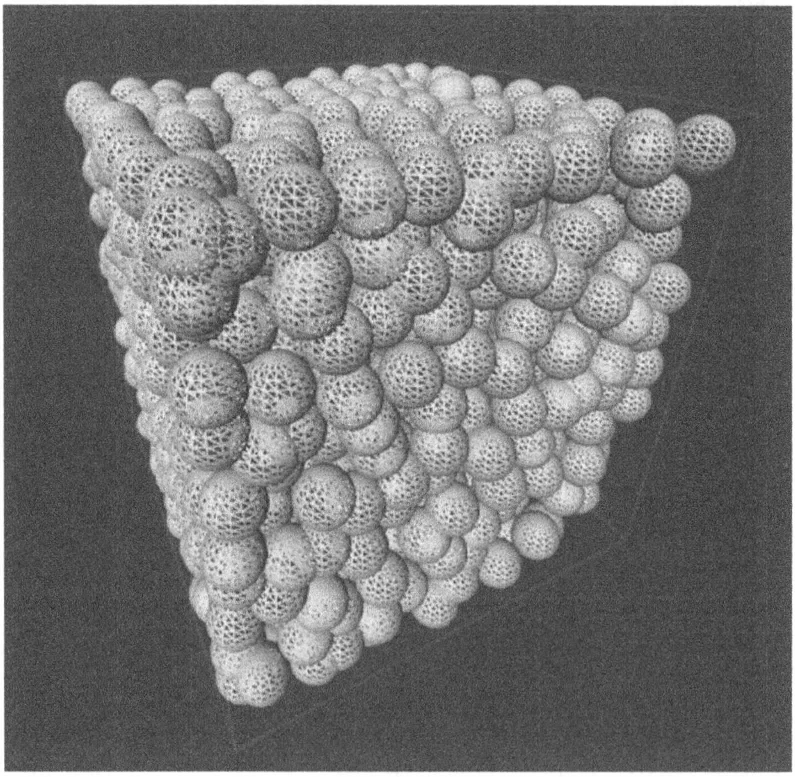

(b)

FIGURE 16.1 a) 170 grains packed in the silo using the above algorithm in 2D, b) 1000 spheres packed in 3D (both done with Prompt 16).

Task 1. Generate a two-dimensional version to pack other geometric figures, for example, squares with random rotation or others. Include physical interactions and moments of force during contact.

Task 2. Generalize the three-dimensional version to packing tori (consider moments of force during contact and the possibility of tori topologically entangling and passing through each other's holes.

Task 3. Based on the simulation, make a simple game in the two-dimensional version. In the simulation, the right edge is movable and "presses the system" by exerting a force on it. The game consists of removing disks selected by the user using the mouse. The user can select a particular disk and remove it. The player loses when the removed disk causes the layout to collapse and move the right edge significantly (e.g., more than a few disk diameters).

17 Double Pendulum

The double pendulum is an interesting physical system in which one rod, hooked in space, is deflected by some angle θ1 like an ordinary single pendulum. However, in addition, there is a second pendulum attached to the end of the first pendulum, which is deflected by an angle θ2 from its equilibrium position (see Figure 17.1).

Under the influence of gravity, the system will move toward a state of equilibrium in which both angles are zero (free dangling of the pendulum). Despite the trivial state in equilibrium, the very arrival at this equilibrium state is interesting, and such a pendulum system has several interesting properties. It does not have an analytical solution. Its motion will therefore be generated using a numerical solution of the equations of motion, which will provide information about the time dependence of the angles and angular velocities. This is because knowledge of the two angles and their velocities is sufficient to completely describe the current state of the system (generalized variables), and from the two angles, we can calculate the positions of the two pendulums and, for example, draw them.

We will take the equations of motion for the two angles from (D'Alessio 2022) (see equations 6 and 7 in there) and write them in simpler form for pendulums with the same masses at the ends and the same lengths. These will be the formulas for their angular accelerations θ1" and θ2" (we will use the ' to denote the first derivative after time, and " to denote the second derivative), which we will write for the angular accelerations θ1 and θ2, respectively:

1) [1 + sin^2(θ2'-θ1')] θ1" – sin(θ2'-θ1')cos(θ2'-θ1')θ1'^2 + (g/l) sinθ1 = sin(θ2-θ1) * (θ2'^2 + (g/l)cosθ2)
2) θ2"+(g/l)sinθ2 = –[sin(θ2-θ1)θ1'^2 + cos(θ2-θ1)*θ1"]

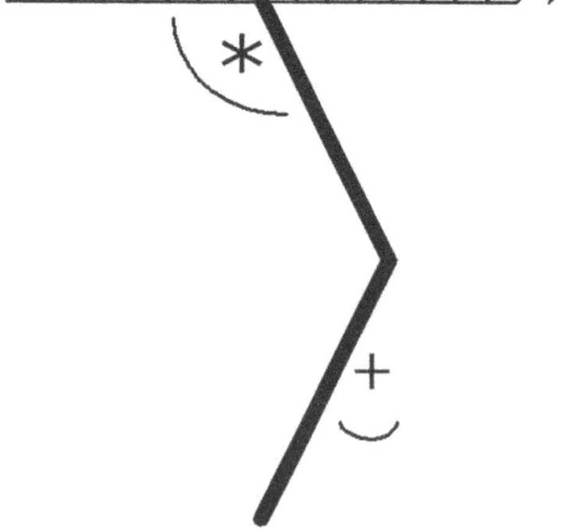

FIGURE 17.1 Schematic of the construction of a double pendulum composed of two segments (thick broken line) with two generalized variables that vary with time (angle θ1 marked * and θ2 marked +).

DOI: 10.1201/9781003734062-18

We can solve the above equations by determining the accelerations $\theta1''$ and $\theta2''$ from the first and second equations, respectively, and integrating these equations, for example, using the explicit Runge-Kutta or Verlet method. We will choose Verlet's method, as it conserves energy.

Let us write down schematically the algorithm of how such a simulation works:

Step 1. Initialize the variables: angles and velocities for the double pendulum (4 variables).

Step 2. Integrate the following equations of motion of the double pendulum to track the motion. Use the Runge-Kutta method 4. Equations of motion for both angles (here the above equations for $\theta1''$ and $\theta2''$).

Step 3. Visualize the pendulums as two segments, one black and one white. You can also draw, for example, how the angles change relative to the angular velocity (phase space trajectory).

We will now use the above algorithm (Prompt 17) and create the code. The results of the simulation run are shown in Figures 17.2 and 17.3.

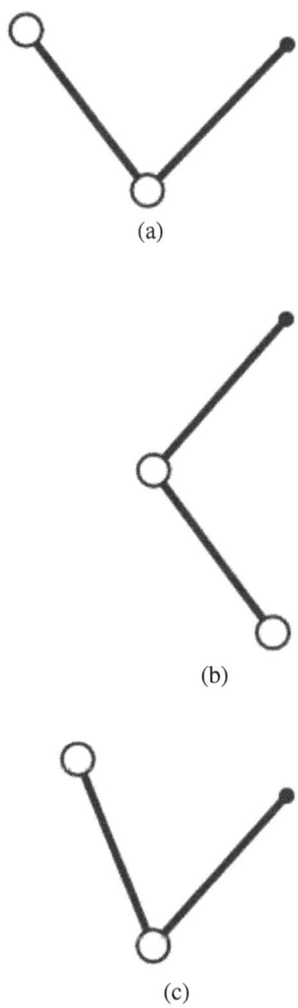

FIGURE 17.2 A double pendulum in a gravitational field (three configurations from simulation). Black circle is fixed; white open circles are moving.

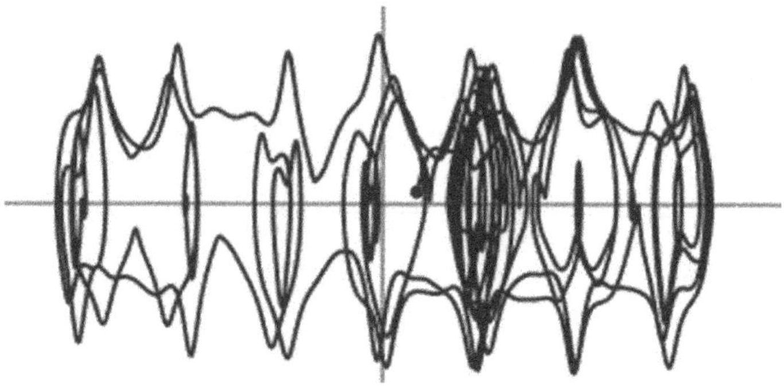

FIGURE 17.3 Phase diagram of the second pendulum, i.e., trajectory θ2(θ2'), where on the x-axis is the inclination angle of the second pendulum segment and on the y-axis is its angular velocity. It can be seen that the trajectory passes quite far in phase space due to the pendulum's rotation around the connection point of the two pendulums.

Task 1. Check the principle of energy conservation in your pendulum. To do this, take into account both the kinetic and potential energy of your pendulum. If energy is not conserved, consider why and whether you are using a conservative method of integrating equations.

Task 2. Simulate two double pendulums that are positioned vertically upwards (maximum deflection of both segments). In one of them, slightly disturb the angle and perform a simulation, calculate the distance between both pendulums in phase space (distance defined as the sum of the distances of the angles and velocities in sequence). Find the Lyapunov exponent for this process.

Task 3. For the exact simulation as in Task 2, plot (visualize) the so-called resonance matrix, in which element i, j contains a floating-point value representing the distance in phase space from the state at step i to the state at step j. By definition, such a matrix is symmetric and has zero on the diagonal. Try to determine, based on the visualization, in which step chaos occurs.

18 Chaos in Double Pendulums

The phenomenon of chaos is related to a physical system's sensitivity to small perturbations. A slight change in the initial condition can significantly alter the system's trajectory in phase space. This is no different in the case of the double pendulum introduced earlier. Deviating it by a fraction of an angle differently (regardless of which angle of the pendulum we are talking about) will cause the ends of the pendulums to have completely different trajectories after only a short while. The difference between the angles will grow over time, and such a distance between two solutions is often measured. In the case of chaotic systems, exponential growth is often investigated (see Lyapunov theory and Lyapunov exponent). This phenomenon can be illustrated by imagining many such pendulums, where each has unique boundary conditions (e.g., angle). However, the initial differences are so minor that they are not immediately noticeable. If such a system runs, after only a short while, each pendulum will find its own path and those paths will separate in the phase space of angles and angular accelerations.

Using the knowledge we already have of one double pendulum, let us try to look at a hundred of them with different boundary conditions, in which we modify the double pendulum algorithm so that the problem is solved for the whole group of pendulums at once. The algorithm is like a single pendulum with change in initialization:

Step 1. Create angle and angular acceleration variables for 100 double pendulums with the same masses and lengths, in which a user-specified angle alpha1 and alpha2 deflect each. However, a random number in the range 0.0001*alpha1 is added to alpha1.

Step 2. Integrate the following equations of motion of the double pendulum so that the motion is tracked. Use the Semi-Implicit Euler method and solve the equations of motion for both angles for each pendulum separately:

for angle 1) [1 + sin^2(θ2'-θ1')] θ1" – sin(θ2'-θ1')cos(θ2'-θ1')θ1'^2 + (g/l) sinθ1 = sin(θ2-θ1) * (θ2'^2 + (g/l)cosθ2)

for angle 2) θ2"+(g/l)sinθ2 = –[sin(θ2-θ1)θ1'^2 + cos(θ2-θ1)*θ1"], where θ1, θ2 are the yaw angles of the first and second segments, respectively. Denote ' (prim), is the derivative, " is the second derivative, "^" is raising to the power, g is the gravity constant, and l is the length of the segments of the two pendulums.

Determine the accelerations symbolically from the equations and use them to solve using the Semi-Implicit Euler method (do not use any other). Make sure the solution is correct.

Step 3. Visualize all the pendulums as two black lines on a white background.

We use Prompt 18 to generate code for simulation of hundreds of double pendulums and computed distance in phase space (see Figures 18.1 and 18.2).

Task 1. Make an animation in which 1,000,000 pendulums are simulated at once with additive visualization (each pendulum adds its own color) with high transparency and an interesting color map of different hues and shades. The animation should be generated and saved frame by frame and then edited into an animation file due to the long generation time of a single drawing.

Task 2. Calculate the time dependence of the Lyapunov exponent for several different initial conditions, initial swing angles, and the degree of variation in the positions (angles) of the initial pendulums. Is there a limit to the properties of the system for which chaotic behavior appears (or disappears)?

DOI: 10.1201/9781003734062-19

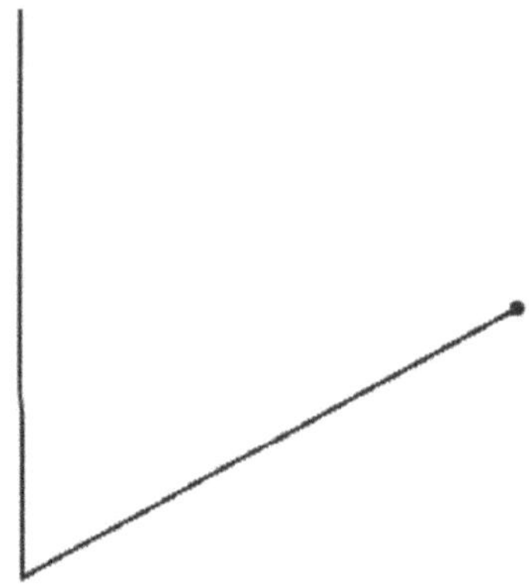

(a)

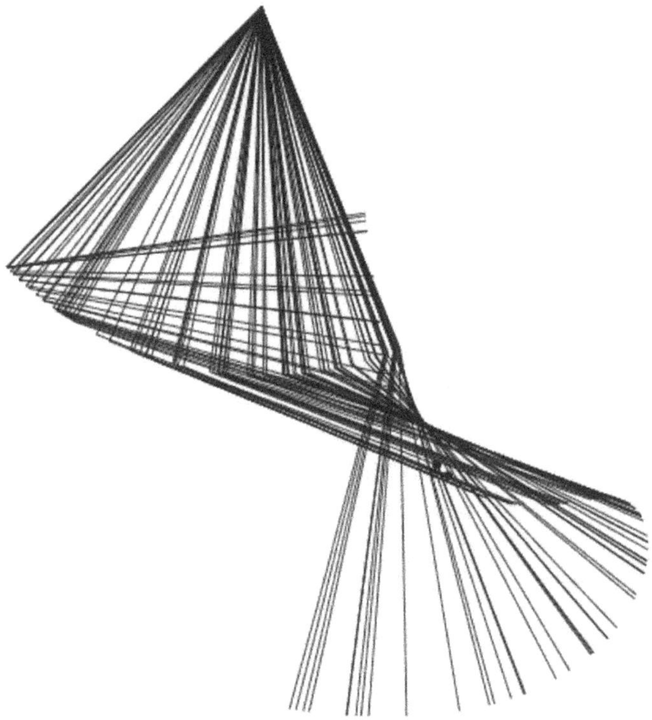

(b)

FIGURE 18.1 A hundred double pendulums with slightly different initial conditions. Simulation snapshots at t = 2.39, 9.33, and 10.54 from (a) to (c).

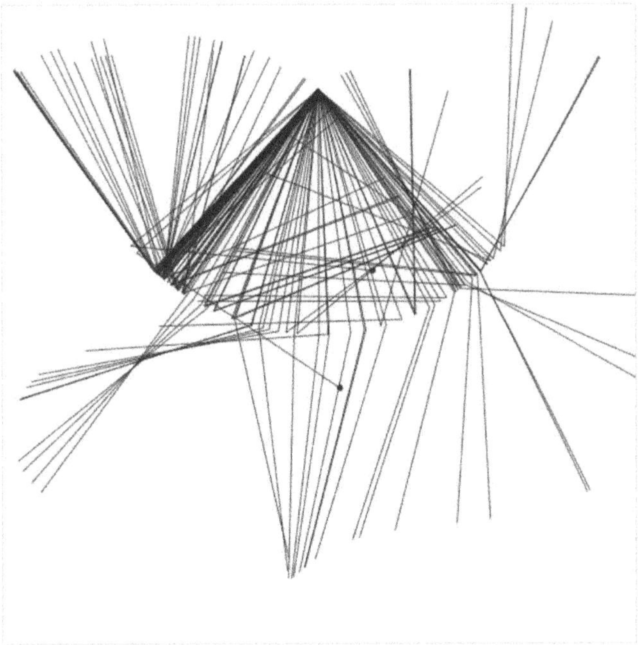

(c)

FIGURE 18.1 (Continued)

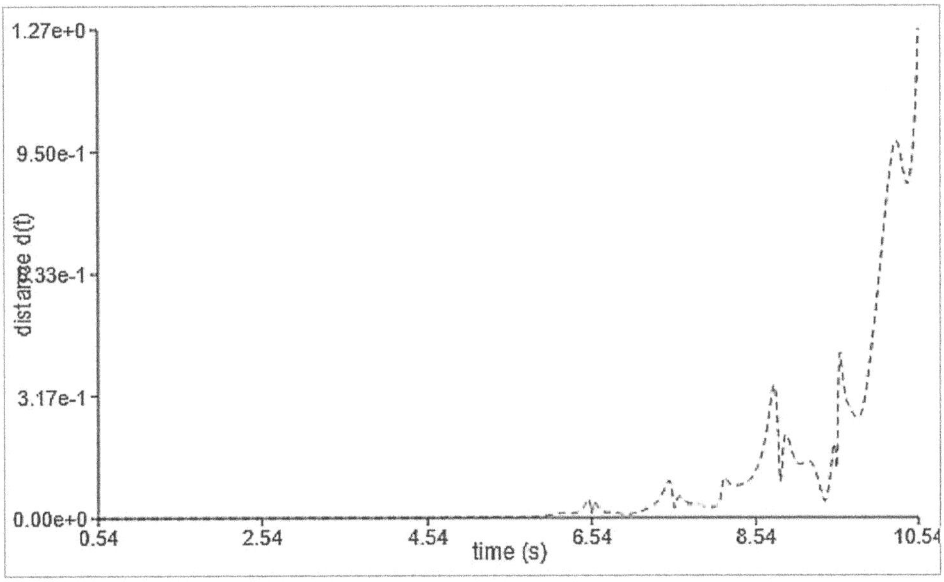

FIGURE 18.2 Distance in phase space between two selected pendulums from simulation of a hundred pendulums. An estimated Lyapunov exponent for this is $\lambda \approx 1.1303\ s^{-1}$.

19 Logistic Map

What is equilibrium in nature? Let us imagine a simple model in which a population (for the sake of illustration, let us assume that it is hares) grows indefinitely according to the simplest possible linear law, which states that a population changes its size linearly over time:

$$x = r * x,$$

where x is the population size and r is a specific coefficient. If r is greater than unity, there are more individuals. For example, for r = 2 and an initial abundance of x = 1, we will have successive values over time: x = 4, x = 8, x = 16, etc. Such a law cannot be fulfilled. A population could only grow in this way if it had an unlimited individual lifespan, an unlimited food supply, and unlimited land to occupy. These conditions will never be met. In 1976, May (1976) described an equation that describes population dynamics much better and is not significantly more complex. We are talking about the logistic mapping and an equation of the form:

$$x = r * x * (1 - x)$$

This equation has an additional term (1–x), which is responsible for the inhibition of population growth (one could say that all the limitations of space, food availability, etc., are hidden here). In this equation, we must assume that the population size has a maximum value of 1, so we will start from some fractional value, and the iterations will resemble those of the previous model. However, it turns out, and this was one of the results presented in May (1976), that this simple model carries several interesting behaviors. For example, what happens for r = 1.5 and abundance x = 0.8? Let us calculate directly from the formula above: x = 1.5*0.8*(1–0.8) = 0.24 (a significant drop in abundance despite the large r). Meanwhile, for small populations (x = 0.2, r = 1.5) we have a different situation: x = 1.5*0.2*(1–0.2) = 0.24, which means that population increased! This means that the system regulates itself and does not allow the number of individuals to fluctuate too significantly. R. May discovered something else, however, as it turned out that for specific values of r the values of x can even oscillate (May 1976). This behavior is not easily predictable, and it is not actually known why it happens. However, it is possible to draw the available values of the final population sizes as a function of the parameter “r” in the above equation. Such a graph is called a bifurcation diagram. Let us write an algorithm that implements the simulation according to the discussed model and generates a bifurcation diagram:

Step 1. In a loop, run through the values of r from 0 to 4.0 (allow the user to select the range of the loop after r, the step of this loop, reset and start the calculation from the beginning) and execute:

- **Step 1.2.** In an inner loop, execute 1000 iterations of x = r*(1 – x), starting at x = 0.5.
- **Step 1.3.** After finishing, perform next 10 iterations and draw values of x on the graph of x(r).

Step 2. Update the graph and make another run of the loop over r. Let the graph show all the calculated final values of x.

Figure 19.1 shows the result of a program generated by the above algorithm using Prompt 19. To go beyond what is usually shown in logistic map studies, we present the behavior for a broader range of the parameter r.

 DOI: 10.1201/9781003734062-20

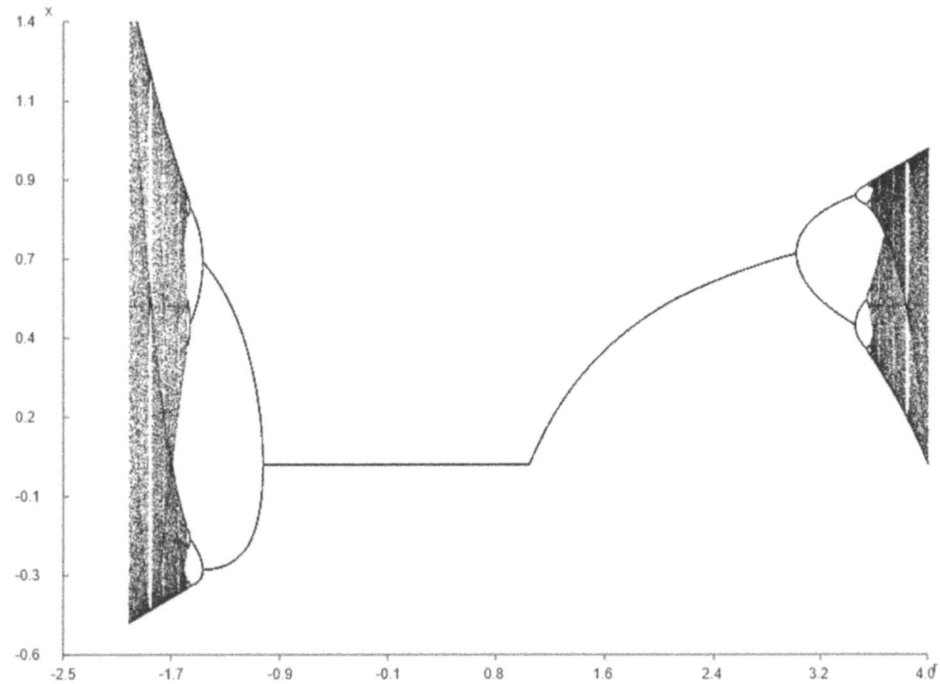

FIGURE 19.1 Plot of x(r) of the 10 final x values calculated after 1000 iterations of the logistic equation x=r*x*(1–x) over the range r from –2.0 to 4.0 of the parameter r according to the presented algorithm.

The figure illustrates the typical behavior of the model, where, around $r = 2.85$, there is a first separation for positive r and oscillation between the two values. Then, x tends toward four and exhibits additional points of attraction. What happens for $r > 3.5$ is absolute chaos.

Interestingly, the area you do not usually look into ($r < 0$) hides chaos of even greater proportions, as our results show here.

Task 1. Investigate the Lyapunov diagrams for the logistic map. To do this, for each value of r in the entire range under study (e.g., –2 to 4), perform two iterations: one for r, the other for r+dr (where dr is a minimal value). Investigate the behavior of the difference in x values as you iterate and fit an exponential function exp(lambda * t), where t is the iteration number and lambda is the exponent you are looking for. Draw the lambda as a function of r over the entire range under study.

Task 2. Modify the iterated function and find another that also exhibits chaotic behavior.

20 Self-Avoiding Random Walk

Have you ever wondered how to draw a maze? The general rule is that a common corridor should connect the entrance and exit of a maze. One of the algorithms that will enable us to generate such structures is a random walk without self-intersections. Let us imagine a wanderer who makes random movements in any direction (for simplicity, it can be a walk in one of the available main directions – up, down, right, or left). For a free particle, such a trajectory can be in the form of a curve that staggers loops, coinciding with itself, because the particle can return to places where it has already been. A simple assumption – that the particle does not occupy the same place twice during its entire history – results in entirely different behavior of the walk. Such a model is visually appealing, but it also possesses several properties, one of which is the ability to generate the title maze. To do this, of course, we would have to generate a trajectory as long as possible, preferably filling the entire system, and this is not possible due to the blocking of the wanderer. Imagine a wanderer staggers down a letter u and then enters it – he will face an impossible movement, where on one side he has the field from which he came (and cannot go back). The alternative three fields are inaccessible to him, because he was already there. To generate a maze, we will use a simple procedure. In a blocking situation, we will remember the current (and all previous) trajectory and move the particle to another random location. In this way, each subsequent walk will already have a little less space to fill, but the system should fill it. Of course, the longer the individual trajectories, the more interesting the maze to walk. Let us write down the simplest algorithm, which describes the procedure of random wandering in the simplest version:

We work on an 800 × 600 canvas and perform a random walk without self-intersections:

Step 1. Check if the wanderer has at least one field around him that he has not been to yet. In a situation where the wanderer has blocked, and none of the four fields are available for the walk. Remember the trajectory and start a new walk.

Step 2. Perform the random wandering step by randomly selecting one of the available directions.

Step 3. Visualize all trajectories as black paths (lines alone, no points). The lines should come to the edge of the layout (box).

Step 4. Allow the user to start, reset, stop the simulation and set the grid resolution (initially 80 × 60).

The result of the algorithm, as performed using a program generated with Prompt 20, is given in Figure 20.1.

Task 1. For each completely generated trajectory of self-avoiding random walk, instead of drawing a broken curve, draw a smooth spline curve that passes through all vertices (interpolation) to form a single, smooth, differentiable, continuous function. Allow the user to modify the smoothness of this curve.

Task 2. Modify original algorithm to change the number of neighbors to 8 and add the mechanics of bypassing possible intersections of trajectories (because in practice, it turns out that with eight neighbors, you still need to add this condition, because otherwise there may be situations when an intersection of trajectories appears in the form of an x).

Task 3. Perform a random walk without self-intersections in a version in which a single walk is performed, and the algorithm includes a backtracking mechanic in the case where the walker is stuck. In this case, take enough steps backward so that further trajectory generation is possible.

 DOI: 10.1201/9781003734062-21

FIGURE 20.1 Labyrinth made using the technique of multiple independent random walks without self-intersections with trajectories. In this final state, there is no room at all in the layout for further walks. I encourage the reader to play along and walk the layout from the center of the left edge to the center of the right edge. Simulation done at 80 × 60 resolution.

Task 4. Generalize the problem to three spatial dimensions and perform a random walk without self-intersections, taking into account six neighboring points. Fit spline curves to the subsequent steps and draw the trajectory as a 3D object, glued and smooth (generate a 3D object described by the resulting curve) in shaded form with lights and ambient occlusion.

21 Snowflake

How is a snowflake formed? In his famous book on cellular automata, S. Wolfram proposed the simplest of all models of snowflake growth as a cellular automaton in which there is only one principle (Wolfram and Gad-el-Hak 2003). This principle states that a snowflake grows by fusing new cells to existing ones. In other words, if we are working with an array and a seed is present, growth will begin from it. In each step of the simulation, all cells that are adjacent to exactly one neighbor that belongs to the snowflake will grow. However, generating an interesting snowflake with the symmetry found in nature is possible on a triangular grid. Working on such a grid is not difficult. It can be represented in an ordinary two-dimensional array; the only thing to remember is that in a triangular grid, each even row has nodes offset by half the length of the odd rows. This arrangement has one consequence: neighboring nodes must be selected accordingly. It is best to draw it out for yourself on a piece of paper, because now each node has six neighbors (right, left, and four diagonals), and you need to take this into account by indexing each node accordingly.

So let us write down the algorithm that realizes the growth of a snowflake in the simplest version given by S. Wolfram:

1 We start with a 200 × 200 array and visualize it on an 800 × 800 canvas.
2 Start with a clean array and a seed (value 1) in the middle.
3 Define the neighborhood so that the even rows are shifted half a cell to the right of the odd rows; that is, our square array represents a triangular network.
4 Any empty cell that has one of the set of selected values, the number of full neighbors, becomes full (dissolving).
5 Draw the array on the canvas (black color value 1, white color value 0) and continue the simulation. Remember to move the even rows when visualizing the triangular network.

Figure 21.1 shows the result of this algorithm implemented using Prompt 21. For reference, the result of a code written in C++ by the author is given.

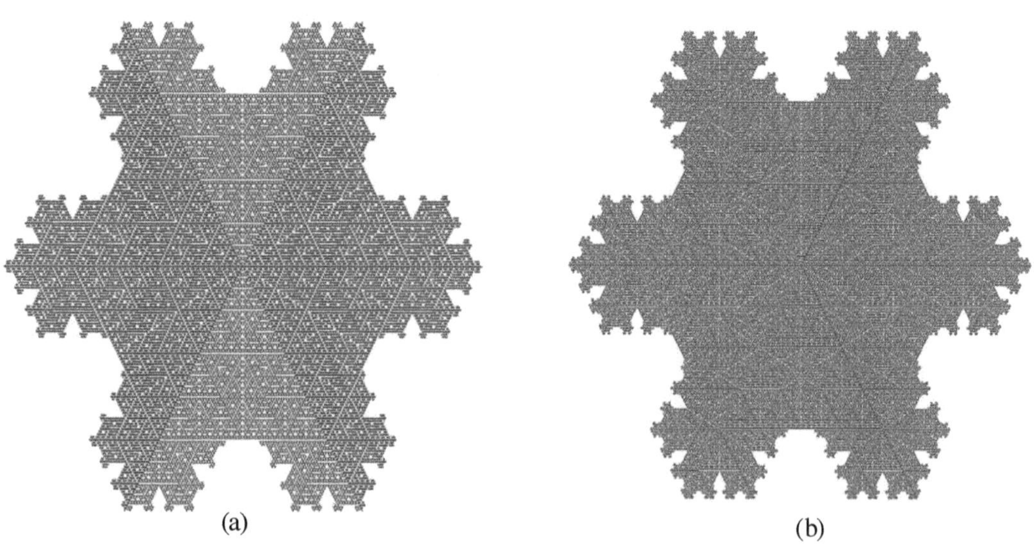

(a) (b)

FIGURE 21.1 Snowflake made using the model and algorithm given above using Prompt 21 a) and C++ code of author for reference, b). Growth rule: 1 neighbor.

DOI: 10.1201/9781003734062-22

Task 1. Consider extended neighborhoods and the associated possibility of having more fully connected neighbors. Let the user freely specify the ranges of the number of full neighbors that cause the cells to grow, as well as specify which neighborhoods we take into account (e.g., using a radius). Try to find some interesting configurations leading to the growth of a broader range of snowflakes.

Task 2. Generalize the model in its simplest form to a 3D version with visualization and growth of an example snowflake.

22 Chaos in Double Numbers

Is it possible to cause chaos by multiplying numbers? It is not hard to guess from the title of this chapter that yes. In his book on chaos, James Gleick proposed a simple algorithm that shows exactly this (Gleick 1996). Let us consider two numbers close in value to each other, for example, x1 = 0.434 and x2 = 0.43400001, and multiply them in a loop by some floating-point number, for example, 1.999.

We will perform this multiplication for a sufficiently long time to observe chaotic behavior. As discussed in the chapter on pendulums, chaotic behavior refers to the phenomenon where two values that are close to each other deviate significantly from one another along a trajectory over time. Here we will observe the values of these two numbers (and their difference) over time. So let us write down the algorithm:

1. a = 0.434 and b = 0.43400001.
2. In a loop (from 1 to 1000), multiply each number in each step by 1.999.
3. If any of the numbers is > 1, then subtract one from it.
4. Draw the graph of a(b).

Figure 22.1 presents a plot of function a(b) resulting from this algorithm implemented using Prompt 22.

FIGURE 22.1 Plot of a(b). One can see the lack of correlation and regularity between the values of a and b, which, after all, started so close and are subjected to the same multiplication operation. Chaos!

DOI: 10.1201/9781003734062-23

Task 1. Make a(b) graph for several different initial distances between the numbers a and b and see if the chaotic nature of the resulting sequences depends on this difference.

Task 2. Draw the graph of a(t) and b(t) and see how they change over time depending on the value of lambda (for several different values).

23 Monte Carlo π Estimation

In this section, we will draw π number. Imagine a circle of radius R, which is inscribed in a square of side 2R. We throw pebbles toward the circle in such a way that they never fly outside the square. There are N pebbles in total. Some of them will fly into the circle (let there be n1 of them). If we perform this procedure for a large number of them, we can quite accurately count the number π from the ratio of the area of the square to the area of the circle. Let us start with two areas:

$$\text{Area of square} = 2R*2R,$$

$$\text{Area of circle} = \pi R*R.$$

If n1 pebbles flew into the circle, and N pebbles flew into the whole square, we can arrange the equation:

$$N1/N = \text{Area of circle/Area of square},$$

which in simple rearrangements gives:

$$n1/N = \pi R*R/2R*2R = \pi/4,$$

and hence $\pi = 4 * n1/N$. Thus, it is enough to throw some pebbles, determine n1, N.

Let us write down the whole process algorithmically:

0 n1=0.
1 Draw in succession N = 10000 pairs of random numbers (x, y) in the range 0–1.
2 For each pair, check that the point (x, y) lies in a circle of radius R = 0.5 with center at the point (0.5, 0.5). If so, add 1 to n1.
3 Visualize the points and compute 4*n1/N estimation on the fly.

The result of executing this algorithm with a program generated by Prompt 23 is shown in Figure 23.1. In addition to visualizing the cloud of points (which we have collected directly based on their positions), we have a graph showing the dependence of the calculated value of π in time.

Task 1. Test several random number generators (LCG, ran0, Mersenne-Twister, etc.) and check the accuracy of approximating π using the above procedure for large N.

Task 2. Propose a procedure for calculating the number π from the volume (points randomized in the arrangement of a sphere immersed in a cube). Check whether the π calculated in this way will be more accurate for the same N compared to the two-dimensional version of the algorithm? Does it depend on N?

DOI: 10.1201/9781003734062-24

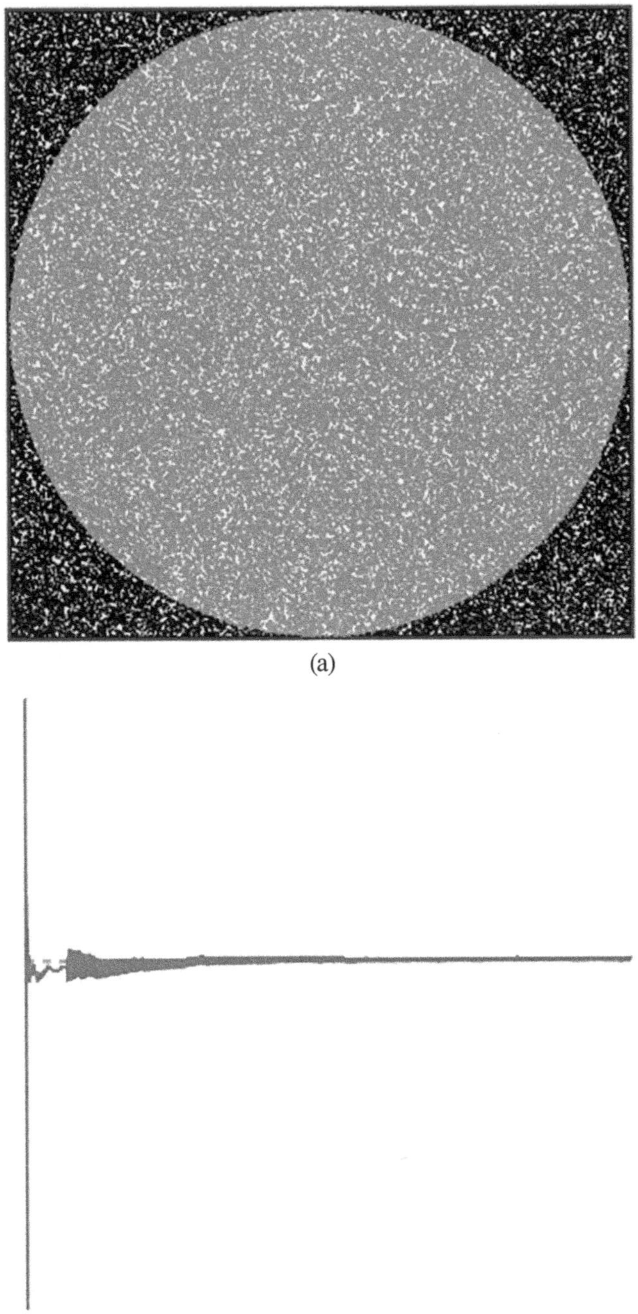

(a)

(b)

FIGURE 23.1 a) Visualization of the simulation – a cloud of 100,000 drawn points, of which 78,466 are inside the circle, which gives $\pi \approx 3.138640$, with error 0.002953 compared to exact value, and b) the graph of the estimated π compared to the book value (horizontal line) in time.

24 Maxwell-Boltzmann Distribution in Gases

Imagine a box containing gas molecules. We know nothing about them except that each of them moves in a rectilinear motion until it collides with some other molecule. If a collision occurs, it will be a perfectly elastic collision. Apart from this collision, there are no other interactions in the system. The system described is one possible representation of the so-called perfect gas. It turns out that if we start the system with random particle velocities and let it reach equilibrium; that is, wait long enough for the system to "relax" and then measure the distribution of the gas molecules' velocities, we will get Maxwell-Boltzmann distribution described by the formula:

$$f(v) = 4\pi\ CD\ v\text{^}2\ \exp(½\ mv\text{^}2\ D),$$

where f(v) is the probability of velocity v, C and D are some constants (e.g., D is inversely proportional to temperature), m is the mass of the molecule, and v is the expected value of the molecule's velocity. Perhaps even more interestingly, if we measure the velocity components separately, both f(vx) and f(vy) will have normal distributions! Let us verify this by performing a simulation using a model of rigid disks that collide perfectly elastically. Let us write down the algorithm for such a model:

1. Distribute the positions and random velocities of N gas molecules as 2D rigid disks of radius R, located in a square of side L. Let R = L/100.
2. For each molecule, calculate the displacement vector at the next time step.
3. Check whether the displacement of any pair of particles will cause them to collide.
4. If so, calculate the time of collision (the exact moment of their contact) and do this for each pair of particles. For each such event, memorize the DT step for which the contact occurs.
5. Sort the array of memorized DT steps.
6. Perform an offset for the smallest of the DT steps, which is the moment of contacting specific disks, and for these disks, perform the elastic collision procedure after this offset.
7. After recalculating the two-body elastic reflection, accordingly, integrate further, taking into account the next DT step from the sorted list to perform the solution of the collision problem of consecutive pairs each time.
8. Visualize the system as a simple black and white drawing of disks in a box in real time.
9. After solving all collisions and performing the entire time step, return to step 2.
10. Live draw a graph (histogram) of the absolute value of the velocity and two additional graphs of the velocity histogram, vx and vy.
11. Allow the user to change N, L, reset, pause, and start.

The results of the above algorithm, calculated using a program created using Prompt 24, are indicated in Figures 24.1 and 24.2. We were able to obtain both normal distributions on the components and a distribution with the shape of the Maxwell-Boltzmann distribution for the velocity modulus.

Task 1. Match the theory – Maxwell-Boltzmann decomposition and Gaussian decomposition – to the distributions you have made and see that the theory describes the results from the simulation well.

DOI: 10.1201/9781003734062-25

Task 2. Perform a simulation for a three-dimensional gas and examine the histograms, both for the absolute value of the velocity and the three components of the velocity.

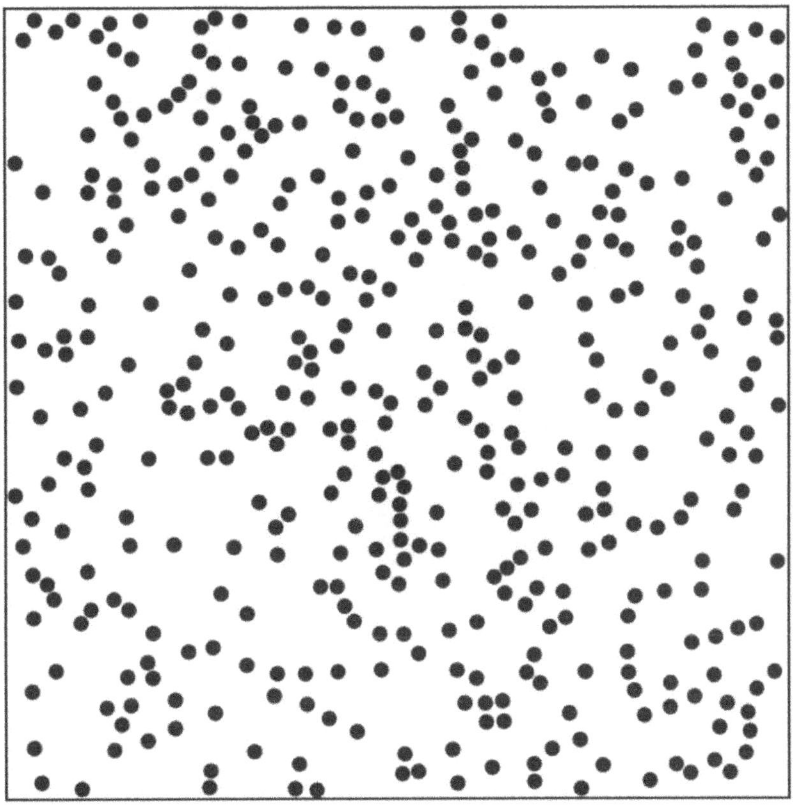

FIGURE 24.1 Configuration of 440 gas molecules (rigid disk model, black circles) moving in a box – here one frame collected from the simulation at the time of collection of velocity histogram plots.

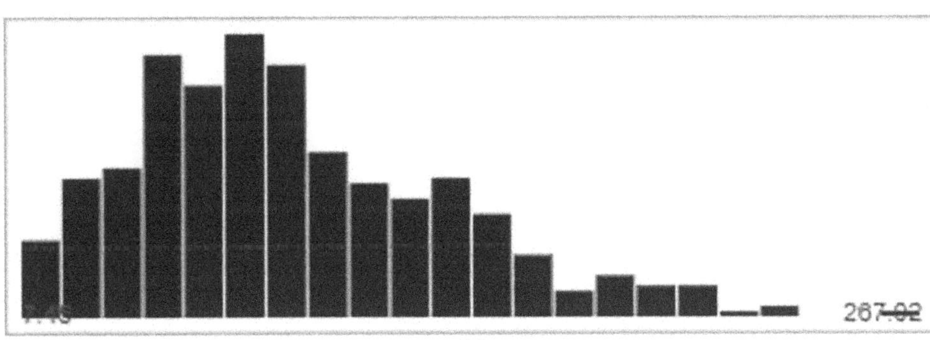

(a)

FIGURE 24.2 a) Histograms of the absolute value of the velocity, b) the vx component, c) and the vy component.

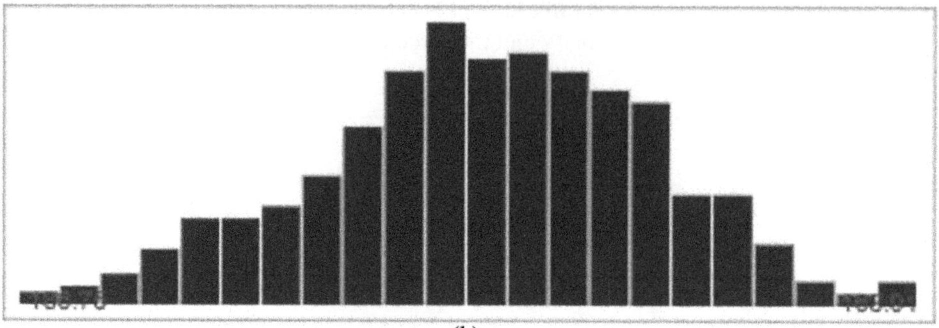

(b)

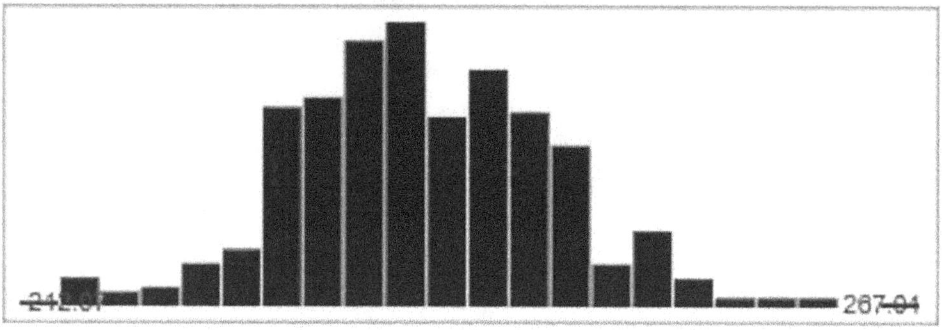

(c)

FIGURE 24.2 (Continued)

25 Forest Fire (Percolation)

When we walk along forest paths, we may come across forest roads – wide, and relatively free of forest litter. One of the functions of such "divisions" of the forest into smaller sections is fire protection. Fire does not easily penetrate long distances. On the other hand, if the forest consists of trees planted very close to each other, a few dry, sunny days and one spark or discarded bottle separate us from tragedy. How to plant a forest to limit the spread of fire? What is the limiting density of trees so that the fire does not quickly get to the other side?

We will answer these questions by building a model of fire propagation in the forest. We will base it on the phenomenon of percolation. Let us imagine that our forest is a two-dimensional array, and the trees are represented in this array (let the empty spaces in the array denote areas without forest litter – inaccessible to fire). To begin with, we will "plant" trees by randomly selecting cells available for tree growth and inserting the value "1" into those cells. The only parameter here is the target tree concentration c, which is a number between 0 and 1. The concentration will be $c = 0$ if the forest has no trees at all, $c = 1$ for a forest filled with trees, and values in between for partial coverage. It is these partial values that will be most interesting. This is because we are interested in the time it takes for a forest fire to spread, calculated from the moment a single tree is ignited on one edge of the system. We expect that for small coverage and c close to 0, the duration of the fire (and the area ignited) will be very small. For a large c close to 1, the duration will be most significant, and the area covered will be close to the forest area. We can also monitor the fraction of lit trees at the end of the fire (that is, the number of lit trees divided by their total number). So let us write the forest fire model in algorithmic form:

Create an array of $N \times N$ with values of 0. $N1 = 0$, $N2 = 0$.

1 Execute a loop after c from values 0.01 to 0.99 with a step of 0.01.
2 Fill the array with one value randomly, with a concentration of c. Calculate the number of ones (N1).
3 Insert the coordinates of all the cells in the first column into the queue.
4 Perform the following steps:
   4.1 Remove the oldest added value (coordinates) from the queue and enter 2 in that place. Add N2=N2+1 (number of lit trees).
   4.2 Perform a loop over the four neighbors of the cell whose coordinates were removed from the queue (for edge cells, take only the available neighbors). If any neighbor has a value of 1, add its coordinates to the queue.
   4.3 If there are any elements further in the queue, go back to 4.1.
5 Draw live graphs of N1(c), N2(c) and (N2/N1)(c).

Allow the user to start, pause, reset, and resize the N grid.

In Figures 25.1 and 25.2, I show the results of the program generated according to the algorithm in Prompt 25. The program calculates both individual fires (color map) and an averaged plot of the fraction of lit trees in the final state of the fire as a function of concentration c (plot). We observe a smooth transition in the model from the system in which the fire is absent to the system in which it is present. This is a percolation phenomenon, and more abrupt behavior can be expected for larger computational grids (with an extreme abrupt jump for an infinite grid).

Task 1. Generalize the percolation problem to three dimensions and study the phase transition, as we did in the two-dimensional case of fire.

DOI: 10.1201/9781003734062-26

(a)

(b)

FIGURE 25.1 End states of fire in an ignited system on the entire column on the left. Simulation on a 250 × 250 grid for initial tree concentration c = 0.58 and c = 0.61, respectively. Light indicates a healthy tree, dark indicates an inflamed tree, and white indicates a space. You can clearly see the phase transition of the system from non-percolating a) to percolating and b) (color figure available online).

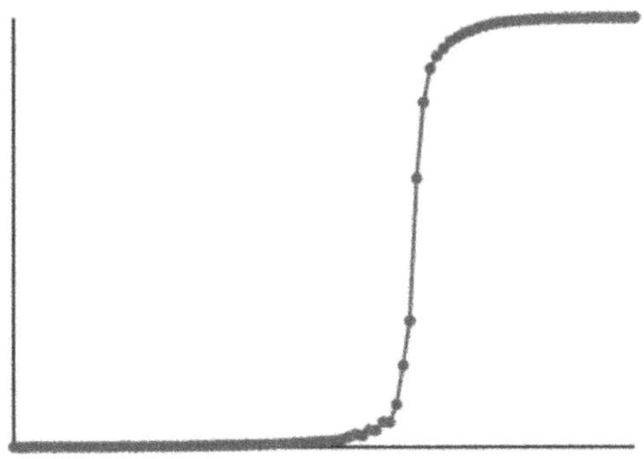

FIGURE 25.2 Fraction of lit trees in the final state as a function of N2/N1(c) for c ranging from 0 to 1 (N2/N1 values also from the same graph). Performed 10 replicates for each concentration of c (mean).

Task 2. Examine the N2/N1 graphs for several grid sizes and determine precisely the value of c for which the graphs intersect. Compare the value obtained with the literature value of the percolation threshold for a quadratic network.

Task 3. Define the model and examine its behavior on a triangular and hexagonal grid.

26 Gravitational Potential

A planetary system maintains its position because gravitational forces act between the planets. The sun, located at its center, has a mass so great that it keeps the entire system stable as the planets orbit around it. From the law of universal gravitation, we can determine the potential energy outside the mass M, which is:

$$P = -GM/r,$$

where M is the mass of the body from which the potential comes, G is the gravitational constant, and r is the distance from the mass under study to the point where we count the potential energy. In the case of a system of several masses, the potential will be the sum of the potential from each mass. It is interesting to study such a potential from a system of masses. If we choose a discrete space to represent our system, we can calculate the potential at each point and draw it so that the highest value is represented by white, for example, and the lowest by black. We will now proceed with a visualization that leads step by step to an interesting three-dimensional representation, where, due to the limitation of drawing on a plane, we will visualize equipotential surfaces. Let us start with an algorithm for a scalar map in 2D:

1 We work on a 1000 × 600 grid, with floating decimal values.
2 Assume that there are 3-point masses in the system (draw their positions and masses)
3 In each cell of the above grid, determine ri (the distance between the center of the cell and mass i) and count: Pi = −G*M/ri, where G = 1, M = 1
4 Calculate the sum of Pi in that cell (contribution from each of the drawn masses) and record it in that cell of the grid.
5 Draw a scalar map (shades of gray) so that the highest value in the array is white and the lowest is black.

The result of executing this algorithm with the program generated by Prompt 26 is presented in Figures 26.1 and 26.2.

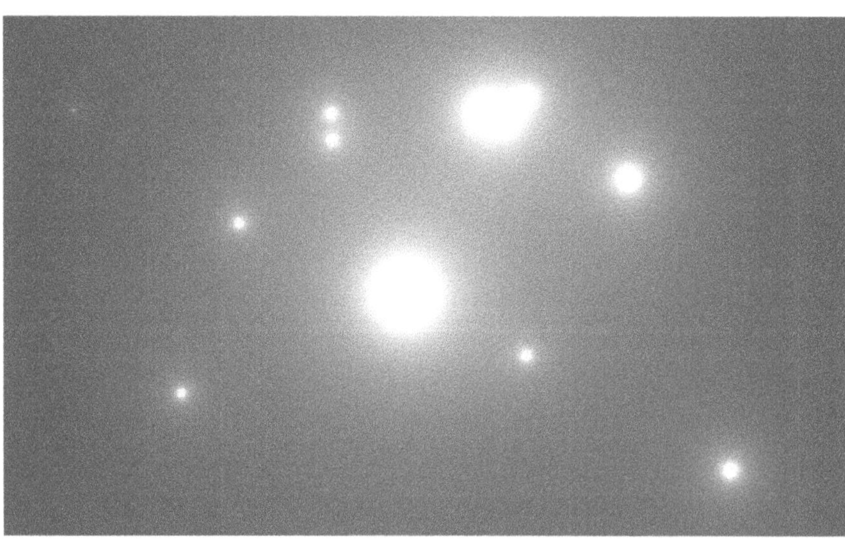

FIGURE 26.1 Gravitational potential of a system of 13 disks with one, the most massive one, placed in the center of the system.

DOI: 10.1201/9781003734062-27

FIGURE 26.2 Gravitational potential of a system of 84 masses arranged in a text.

Task 1. Generalize the calculation to 3D and use the marching cubes algorithm to visualize the isosurface of the resulting scalar field of gravitational interaction. Allow the user to select a value on the isosurface to visualize.

Task 2. Make an animation of the gravitational potential of a solar system-type system with a huge mass at the center and gravitationally interacting planets spinning around it.

27 Sand Cellular Automata

Sand consists of tiny grains that do not stick to each other (well, unless it is wet, but let us deal with dry sand). When such a grain finds itself in a gravitational field, it will begin to fall. When it hits another grain that is below it, it may fall in a random direction, simply bouncing. It can also happen that the grain stops, for example, on the ground and does not fall at all, and the next grains falling in that place will form a mound. Let us save such a model for Chopard and Droz (1998) as a cellular automaton. We will model it in two dimensions, on a square grid, where grains will be represented by one and space by 0. We will decompose the rules for the fall of sand grains into a set of 9 rules that describe how the so-called Margolus Neighborhood changes, which is a 2 × 2 grid with a specific configuration of grains. Let us write them down as two columns, where the 2 × 2 values on the left are the situation at time t, and the 2 × 2 values on the right are the situation at time t+1:

t t+1

10 -> 00

00 10

The first rule states that the seed located in the upper left corner of a 2 × 2 grid, with an available space under it, will fall one grid down in the next step. The next seven situations are already easy to understand by analogy:

01 -> 00

00 01

10 -> 00

10 11

10 -> 00

01 11

01 -> 00

10 11

01 -> 00

01 11

11 -> 10

10 11

11 -> 01

01 11

The final, ninth rule will deal with the situation when two grains, glued together, fall together. In such a situation, in order to model something like friction, or resistance, we will let such

DOI: 10.1201/9781003734062-28

grains fall with a certain probability "p," which will change the nature of the sand falling at high density:

11 00 (with probability p)

00 11

We can write such rules compactly using simple algebraic expressions, but we will leave this method for the reader to study on his own (Chopard and Droz 1998). Instead, let us simply write down the algorithm of the sand fall model using the above information and supplementing it with fields on which the sand rests freely:

1 Consider a 200 × 100 grid, values of 0 and 1, conditions zeroing the values on the boundary cells.
2 Fill the array with random values of 0 or 1 with probability 0.5.
3 Execute an algorithm that in 2 × 2 grids passing through the entire layout for steps even starts from cell 0, 0, and in odd steps from cell 1, 1, performs the following value swapping if the 2 × 2 pattern agrees step t, swapping to the configuration at time t+1 (written here in the form of the 2 × 2 value at time t and next to the 2 × 2 value at the next time step):
(Listed here are rules 1–9 mentioned earlier.)
4 The last row at the bottom should be a blocking row, filled with blocking cells in a strip 1/3 the width of the layout (the sand that falls on it rests).
5 Visualize the values of 0, 1 from the array on a 600 × 300 canvas and continue executing the above algorithm.

The results of this algorithm, written in a program using Prompt 27, are shown in Figure 27.1.

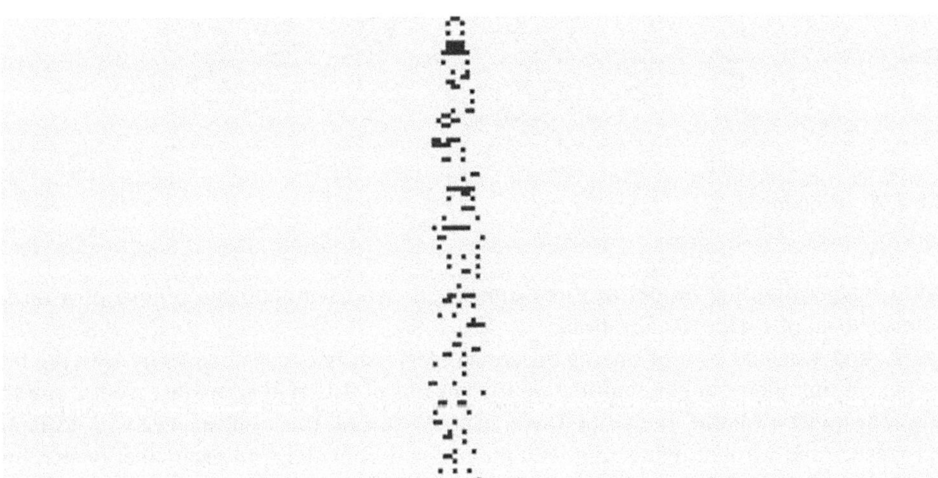

(a)

FIGURE 27.1 a) Free fall of sand from a source, b) Sand falling down on a horizontal surface drawn by the user, and c) complex geometries and sand falling through many holes.

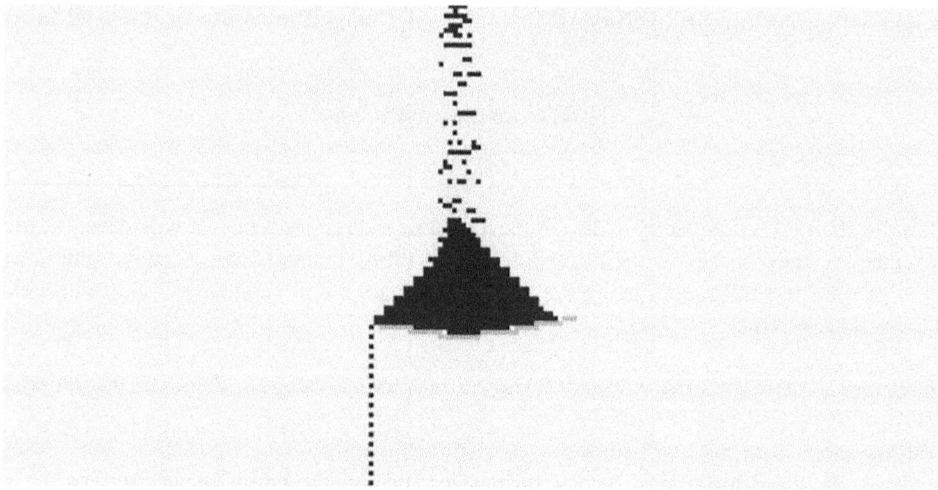

(b)

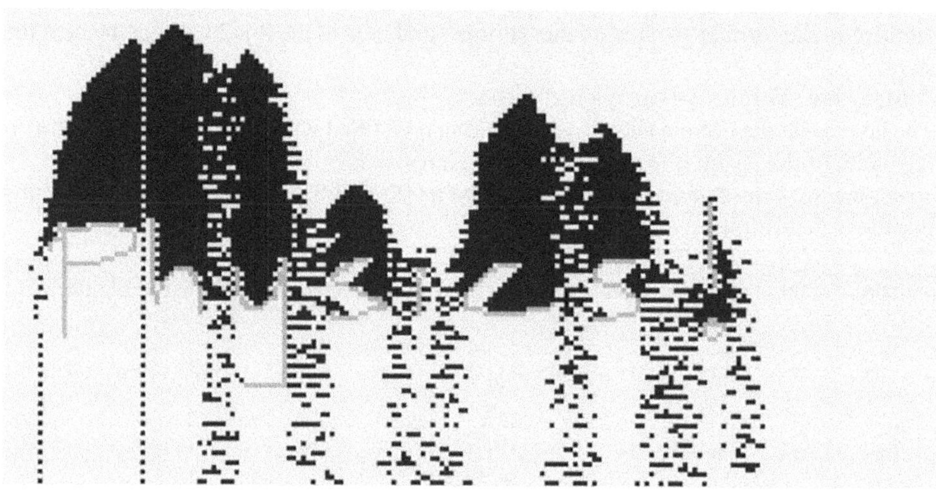

(c)

FIGURE 27.1 (Continued)

Task 1. Generalize the model to three dimensions and simulate the repose of a three-dimensional pile with visualization.

Task 2. Add a second type of sand (a different color) that interacts chemically with the first type, causing their mutual annihilation (two grains of different types that meet cause the disappearance of both). In addition, you can add several other substances with different properties and ways of interacting with the original sand and with each other (water, fire, fuel, etc.).

28 Sierpiński Triangle

The Sierpiński triangle is a self-similar figure (fractal) composed of a specular structure made up of triangles. There is an interesting algorithm for its generation, which we will present in this chapter. If we take an equilateral triangle and randomly select points within it by performing a simple transformation, we should obtain a Sierpiński triangle figure. This transformation involves choosing (also randomly) one of the vertices of the triangle and moving half the distance. After this shift, we draw a vertex in a new position and repeat the procedure. As a result, let us see for ourselves. Let us write down the algorithm using the recipe ("Chaos game") from Wikipedia https://en.wikipedia.org/wiki/Sierpiński_triangle:

1 Determine a triangle with side a.
2 Draw a vertex inside the triangle.
3 Draw one of the vertices.
4 Move half the distance toward the drawn vertex.
5 Draw a vertex and return to point 2.

The result of the program based on this algorithm, written with Prompt 38, is shown in Figures 28.1 and 28.2.

Task 1. Color the triangle by selecting a color palette that maps color brightness to the order in which points are placed.
Task 2. Look for other interesting patterns in polygons with different shift lengths within the algorithm.
Task 3. Generalize the model to 3D by placing figures (for example, spheres, cubes etc.) instead of points, and perform the algorithm for a regular tetrahedron and a cube.

FIGURE 28.1 Sierpiński triangle made using the above algorithm.

DOI: 10.1201/9781003734062-29

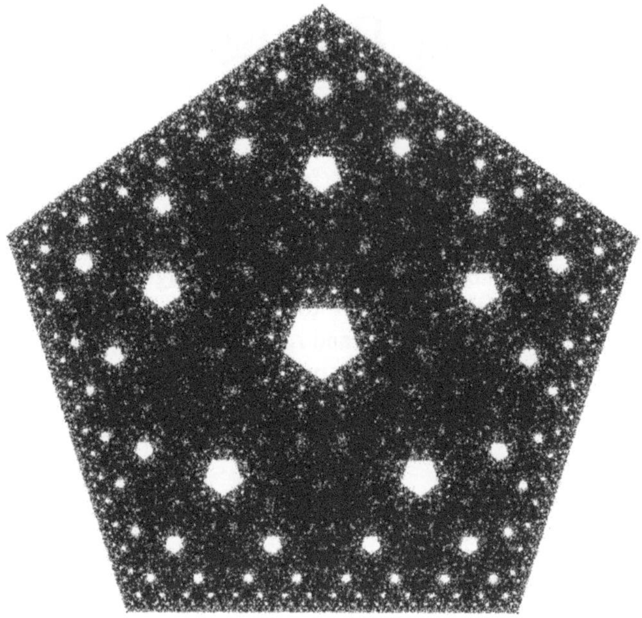

FIGURE 28.2 We can also perform the same algorithm for other polygons and obtain other self-similar figures. Here is the execution of the iterative algorithm for a pentagon.

29 Dense Packing

There are many ways to pack objects in a box. We can put them in one by one, we can stuff them, but we can also do it completely randomly – each time finding a place where such a model will fit. So far, we have considered either a physical model, where growing spheres push up packing in a silo, or a fractal model, where triangles "hole" the space. Now we will try to pack figures of any type, putting them "in the box" randomly, so that they fill the entire space (Baram and Herrmann 2005; https://paulbourke.net/fractals/randomtile/). Moreover, we will make them smaller in each step, so that they will always find a place for themselves, and in the end, we will achieve a really dense arrangement. Let us start with something simple and pack the circles in a square box. The circles will initially have a specific size R0 and will decrease their radius according to a specific function, starting linearly.

Let us write down the algorithm according to which we will be able to perform such "packing":

1 We work on a square area of size L.
2 We define a function g(i) = 1/i^b, where the sign "^" is raising to a power, i is the iteration number, and b is the given exponent.
3 For the given exponent "b," choose R0 so that g(0)*R0<L/6.
4 Increase the iteration count n = n + 1. Calculate R(i) according to the function R(i) = R0 * g(i), where i is the iteration number.
5 Randomly find a place in the layout to insert a circle of radius R(n) so that it neither overlaps another circle existing there nor touches an edge. Insert the circle there. If you cannot find the place in K tries, go to step 3.
6 Draw the layout (all added circles as black circles on a white background).
7 Return to point 3. Continue until the user stops the program.

The result of the algorithm executed by the program written using Prompt 29 in Figures 29.1 and 29.2.

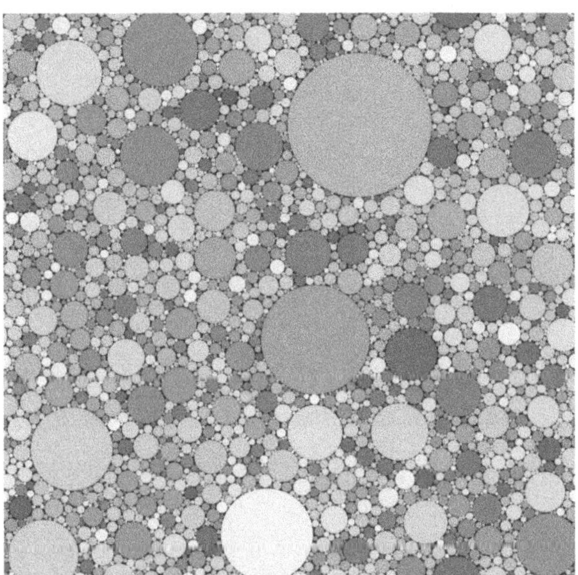

(a)

FIGURE 29.1 Circles and squares rotated by a random angle for decreasing size according to the above algorithm for exponent b = 0.4 and b = 0.3, respectively.

DOI: 10.1201/9781003734062-30

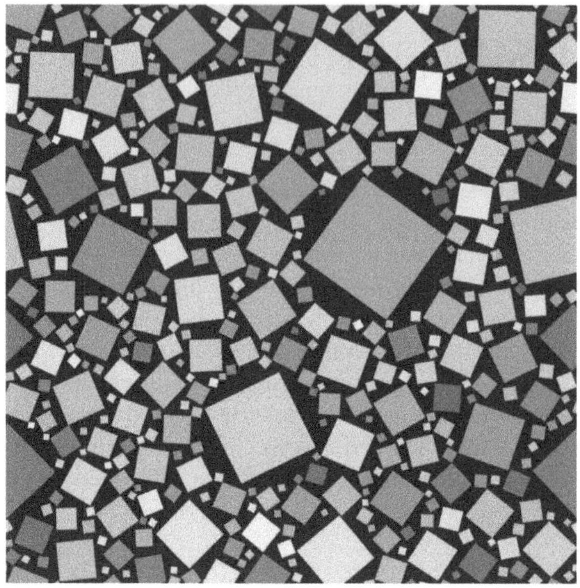

(b)

FIGURE 29.1 (Continued)

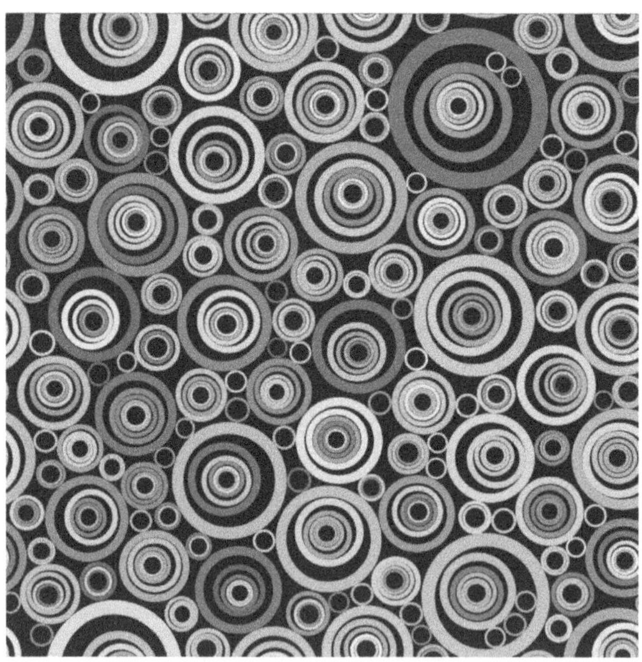

FIGURE 29.2 Packing of tires (bearings) of different radii made with the above algorithm.

Task 1. Generalize the packing program to three dimensions and perform visualization with shadows and global illumination for various three-dimensional objects.

Task 2. Use the box algorithm to calculate the fractal dimension and compare its values for different configurations of packed objects and different exponents in the scaling function of the objects.

30 Lissajous Figures

If we deviate a physical pendulum, suspended in a gravitational field, from its equilibrium position, it will begin to swing around the equilibrium position. If the tip of such a pendulum were to draw a projection of its position on a surface, it would outline the surface in various types of patterns. These can be circles (which will be formed when the deflections in both directions are the same), or other, more complicated patterns. These patterns are called Lissajous figures (https://en.wikipedia.org/wiki/Lissajous_curve). Often, such an experiment is performed by suspending a container of sand from the end of a pendulum. This one, piled up, draws patterns, leaving a trace of its movement. Let us write down an algorithm to generate Lissajous figures by solving the pendulum's motion in 3D:

1 The solution of the pendulum's motion is determined by the formula for the pendulum's deflection in time: f(t) = f0sin(w0t + fi), where f0 is the amplitude of the oscillation, w0 is its frequency, and fi determines the initial phase of the oscillation.
2 Determine independently the deflection of the three-dimensional pendulum in time separately for the x-axis and y-axis using the formula from step 1.
3 For the given initial conditions and f0, w0, and fi for both directions of pendulum deflection, calculate the position of the pendulum at time t.
4 Draw a point at the determined position on the plane.
5 Continue the loop in time by drawing successive pendulum positions (connect the current point where the pendulum is with the previous one).
6 Allow the user to change the initial parameters of the pendulum swing and the rest of its parameters, change the size of the canvas, clean, reset, pause, and start the simulation.

The program, following the given algorithm (here made with Prompt 30), can generate figures for pendulums with arbitrary parameters, such as those shown in Figure 30.1.

Task 1. Generalize the graphs to three dimensions and draw configurations for different parameters. Try to obtain closed loops and three-dimensional figures similar to those in Figure 30.1.

Task 2. Simulate a damped pendulum and draw Lissajous figures for different damping parameter values. Try to create an impressive visualization, for example, by drawing successive simulations on the same canvas with decreasing damping and increasing transparency.

DOI: 10.1201/9781003734062-31

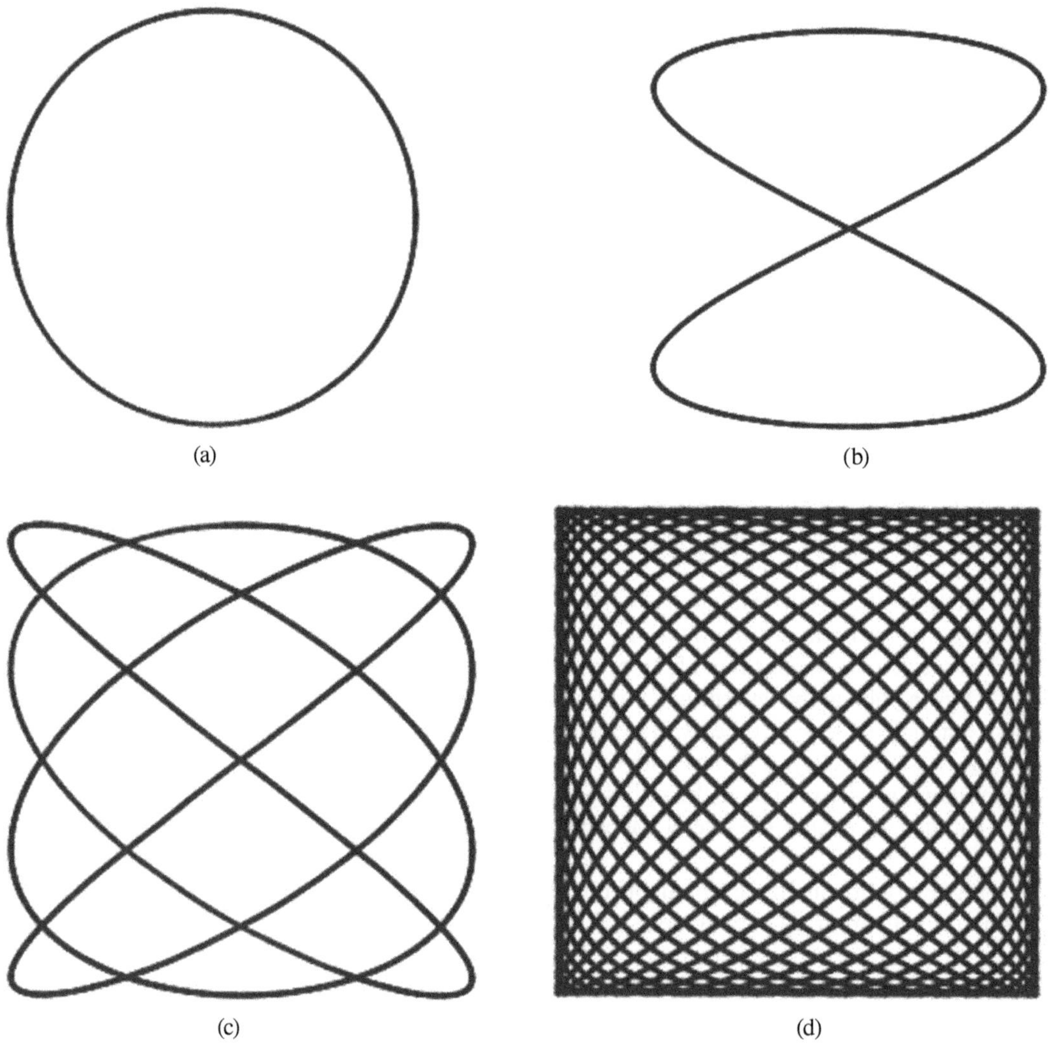

FIGURE 30.1 Lissajous figures produced by pendulum motion in two planes for different parameters (from left): two equal frequencies, one of the frequencies is doubled, the frequencies are close to each other, but not equal, and (last figure on the right) frequencies are even closer.

31 Off-Grid Bacterial Colonies

Bacterial clusters formed in nature are not built on a grid, as we showed before. In fact, they are in free space and can cluster more freely. In 1995, a paper was published on Eden's continuous model, which is designed to model the growth of bacterial clusters composed of attached circular objects representing bacteria (Wang et al. 1995). The model is a simple concept. It involves setting up the first cell (embryo) on a canvas. With such a germ, another bacterium can form, which attaches to it at a random location by coming into contact with its surface. After such attachment, a cluster of two cells is formed. In the next step, we add another cell in the vicinity of this cluster, forming a layer around the joined cells. The distance from the center of the bacillus is used as the radius, and a random location is selected for attaching another bacterium. We continue this procedure until the desired size of the bacterial colony is reached.

Let us write down this growth algorithm in algorithmic form:

1 We work in a continuous space (in a box) of side L
2 In the center, we place a live bacterium (circle) of size R = L/128
3 From the living bacteria, draw one. Around it, draw a possible space (a free space where there are no other bacteria and where a bacterium at a distance R can be attached). To do this, sample its surroundings by selecting 360/K random angles from 0 to 360 degrees and check if another bacterium can be attached there (if the spot is empty). If there is no space available around the bacterium in question, it dies, and a new bacterium is chosen randomly from the live ones.
4 Attach a bacterium to the place found in point 3.
5 Continue the loop and add more bacteria.
6 Visualize the arrangement in two ways: the first, white circles with a black border on a white background, with no other elements or colors; the second, a growth map, showing all growth events as lines that connect the cell around which the next one appeared with the center of this added one – here draw black lines alone on a white background

Figures 31.1, 31.2, and 31.3 visualize the growth of bacterial colonies generated using the algorithm described earlier, with a program created using Prompt 31. The algorithm was extended to include non-circular shapes of bacteria, as indicated in the last point (results are also given next).

Task 1. Generalize the code to three dimensions and draw (visualize) the growth of bacterial colonies composed of bacteria in the shape of small spheres (spheres) and cylinders (sticks).

Task 2. Calculate the dependence of the radius of the cluster (calculated as the radius of the circle described on the cluster with the center at the center of gravity of the cluster) on the number of bacteria that form it, and show this graph for several ratios a/b in the case of a model of elongated bacteria with length a and width b.

DOI: 10.1201/9781003734062-32

FIGURE 31.1 Resulting cluster in the continuous (off grid) Eden model for bacterial growth done using presented algorithm.

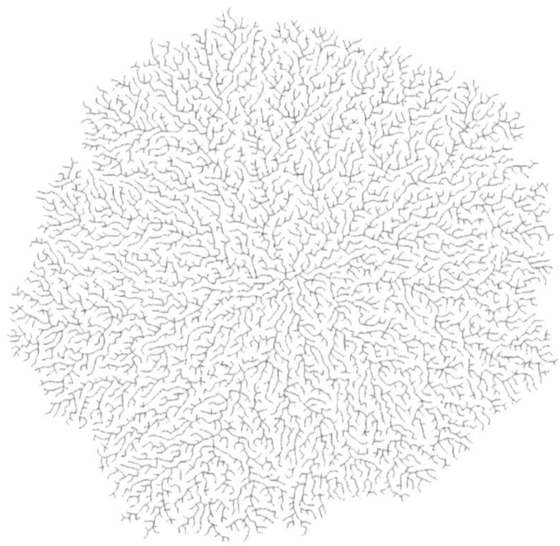

FIGURE 31.2 Cluster growth network from the previous figure.

FIGURE 31.3 Growth of a bacillary cluster composed of bacteria with an elongated shape.

32 Spring-Mass System (Soda Constructor)

A long time ago, in the depths of Internet history, when Windows 95 or maybe 98 icons adorned the wallpaper and Java applets were leading the way on the Internet, Soda Play created a program called Soda Constructor, which simulated objects composed of points connected by springs that mimicked the action of muscles. The program achieved great popularity; it was unique, based on a simple model of springs, and allowed users to build their own complex objects that could move on their own accord. A whole community was formed to create such objects, online forums where people exchanged their objects. A colleague introduced me to this program, and it was probably at that moment that physics revealed its other, computer-based face. The basis of the Soda Constructor program was Hooke's law describing the dependence of force on the deflection of a spring from its equilibrium position says that the force from the extension or compression of a spring is directly proportional to the deflection; that is: $F = -k*(x-d)$, where "x" is the current length of the spring and d is its equilibrium position. In this case, if x equals d, equilibrium will mean the absence of elastic force. If we connect two masses with a spring, they will interact with each other; one mass will exert a force on the other, and vice versa. The spring will strive for the equilibrium position by vibrating around it. With such simple connections, we can build more complicated models, where the simplest, yet stable, structure is a triangle made of three masses connected by springs along its edges. Such a triangle can even be treated as a larger model, placed in a gravitational field, and allowed to fall onto a surface from which the masses that make it up will bounce as its vertices. This is how a simple solid model is created, which we will also attempt to animate by introducing muscles. If we make the equilibrium position dependent on time, that is, substitute the function $d = d(t)$, the spring will react as if a muscle were attached to it. Depending on the form of the function d(t), the behavior will be different. We will assume harmonic motion, consistent with the sin(t) function. Choosing the arrangement of such "muscles" appropriately and performing the simulation in an appropriate physical configuration, such as an arrangement of masses with springs and muscles in a gravitational field, allows us to put complex objects into motion, including robots, vehicles, and characters. Let us write down an algorithm that will realize such a simulation. Let us add to it elements that allow us to add new masses, set muscles, their parameters, and build complex figures.

Let us write down the general algorithm of such a program:

1 Conduct a numerical simulation of a system of point masses with gravitational and Hooke's forces.
   1a Let the Hooke's force be linear, acting in 2D space in a vector acting in the appropriate direction on the points, and let it take into account the damping of the spring linear with respect to the difference in velocity of the two points attached to it.
2 Use Verlet's algorithm to integrate the motion of the masses.
3 Allow the user to add points (masses) by using the mouse.
4 Allow the user to add an elastic connection by clicking on a point, dragging the mouse, and pressing the button over another point (the current length becomes the equilibrium position). Show the connection during such mouse movement.
5 Simulate on a canvas of size W × H with a floor at height H/10 from which material points bounce with damping with coefficient $f = 0.8$.

DOI: 10.1201/9781003734062-33

6 Allow the user to change (by right-clicking on a given spring) the type of spring from an ordinary spring with a fixed equilibrium position to a muscle-type spring whose equilibrium position changes according to the formula d(t) = A0sin(wt+fi).
7 Allow the user to change A0, w, and fi for each spring after it is selected if it is a muscle.
8 Allow the user to change W, H, f, gravity, and integration step dt.
9 Allow reset, clean, pause, start.
10 Visualize straight, in 2D, black lines representing springs, dashed if the spring is a muscle, minor points as masses.

The program implementing the simulation with this algorithm was generated using Prompt 32 (see results in Figures 32.1–32.3).

Task 1. Create a procedure that randomly creates "robots that walk" and optimizes their shape and structure to walk as far as possible without stopping.
Task 2. Generalize the program to three dimensions and build an object that can move on a surface.

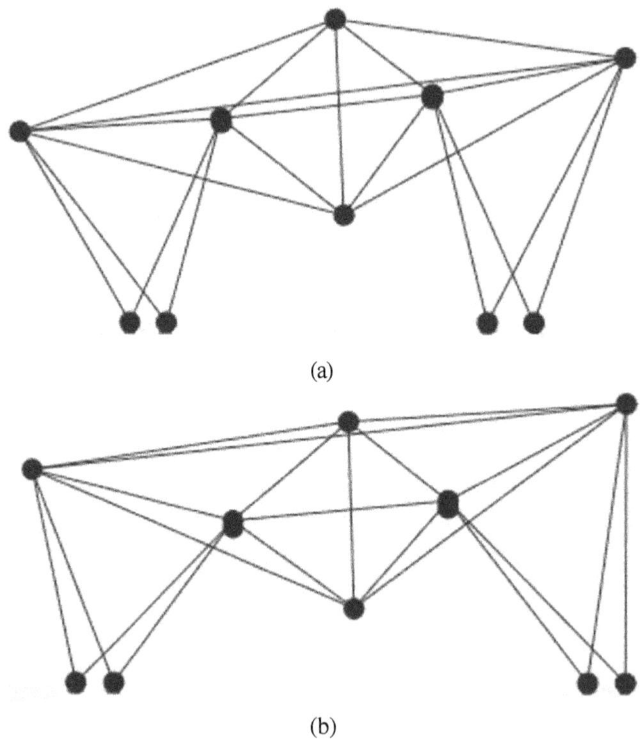

FIGURE 32.1 The famous "Dainty Walker" model from the original Soda Constructor application – here simulated with our algorithm and tool. The situation represents increasing gravity.

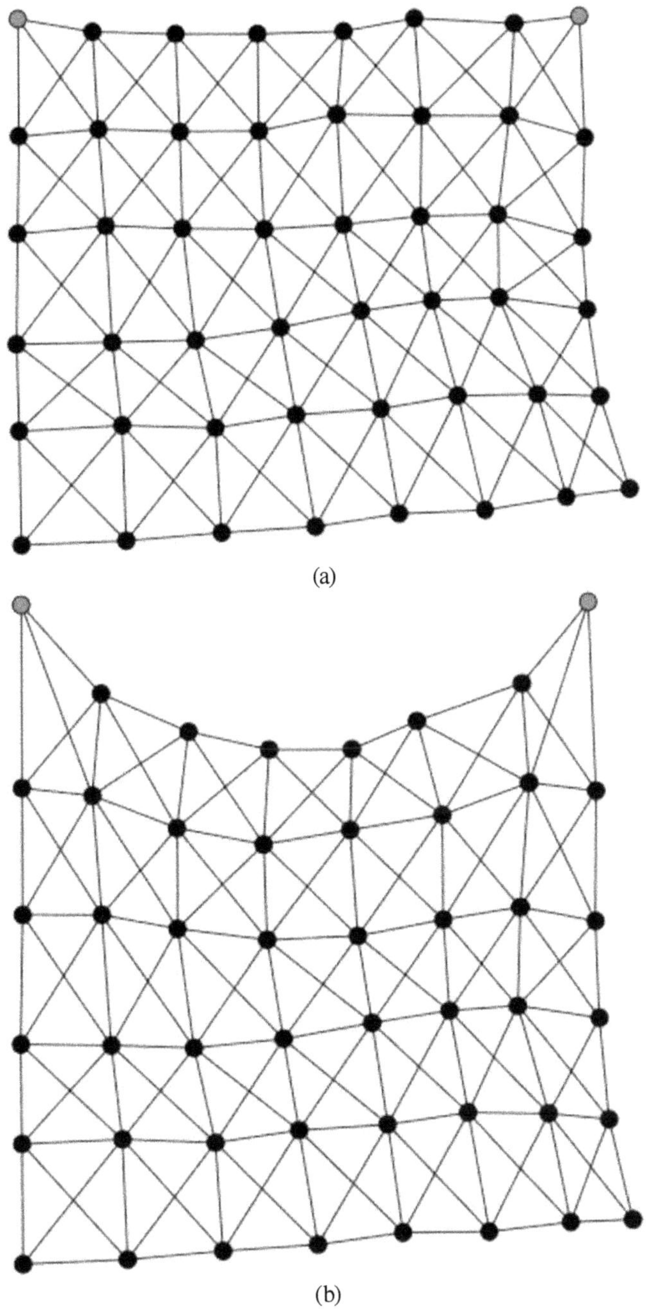

FIGURE 32.2 A model of fabric in a gravitational field suspended by the corners.

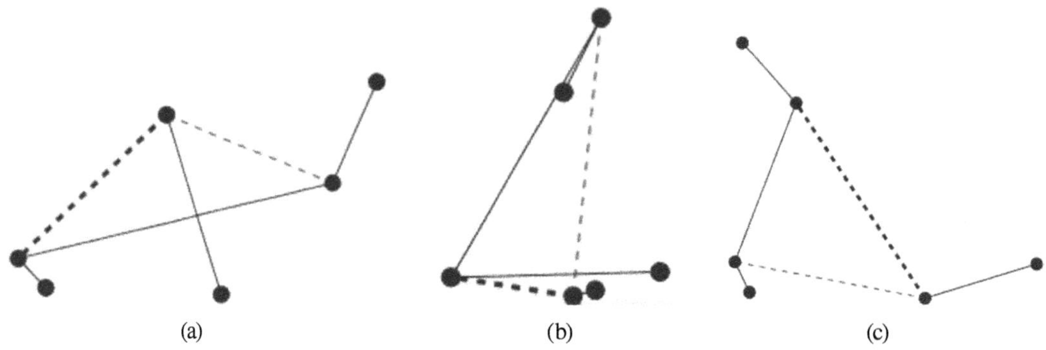

FIGURE 32.3 Triangle with two muscles jumps around and moves due to the muscle's activity.

33 Soft Body

A soccer ball, a balloon, and a bicycle inner tube – all these objects share some feature. They are flexible membranes (materials) that seal tightly and contain a gas pumped into them at a relatively high pressure. The pressure must be chosen with a head (in simulation, too), because it must be large enough to maintain the body's shape and small enough not to cause it to burst. In this chapter, we will simulate such a body based on Matyka and Ollila (2003) and classical mechanics and the model of masses suspended on springs introduced earlier in this book. To achieve this, we will first create a skeleton of the object, whose surface will consist of points connected by springs. Such a model, however, will not behave well – dropped on a surface, for example, it will hit the surface and collapse. To maintain its volume, the model requires an additional force, which is generated by the pressure difference between the gas in the atmosphere and the gas inside. Such a force will be proportional to the inverse of the volume of the body (the area of the surface, if we do the simulation in two dimensions); that is,

$$P = C * 1/V,$$

where P will be the value of the force, C will be a constant (chosen to get the desired effect), and V will be the volume (area) of our body. Such a force should act on each element of the surface along its normal vector – exactly as the pressure force (by definition, the force exerted on the surface) acts.

After overcorrection of the equations of motion of the points on such a membrane, taking into account the aforementioned pressure force, we should obtain a model of, for example, a ball or balloon that maintains its volume. Due to the complex physical model and the large forces involved, these models deform significantly when subjected to considerable pressure. Therefore, we will additionally impose the so-called inverse kinematics constraints (Provot 1995). They consist of limiting the stretching and compression of the spring so that its deformation does not exceed a small percentage of its length at equilibrium. The complete soft-body model is described in more detail in (Matyka 2021), among others.

Here we will give a complete algorithm to simulate it:

1 Conduct a numerical simulation of the motion of point masses in a gravitational field with springs (Hooke's forces with damping) and with pressure.
  1a Simulation on a canvas of size W × H with a floor at height H/10 from which material points bounce with damping with coefficient f = 0.8.
  1b The initial condition is N points uniformly distributed on the edge of a circle of radius R. Each point is connected by a spring to its neighbor (a closed loop is formed).
2 Use Verlet's algorithm to integrate the equations of motion of the points.
4 Allow the user to move the points during the simulation (by grabbing with the mouse).
5 At each time step, calculate:
  - gravitational force (for each point)
  - Hooke's elastic force with elasticity k and motion damping constant kd – for each spring connecting two points
  - S – the area of the body (approximated by the rectangle described on it, the area of which will approximate the area of the whole body)

DOI: 10.1201/9781003734062-34

6 In each step, count the pressure as P=C * 1/S, where C is a constant and S is the approximate volume of the body.
7 The vector value of the pressure force acting on each point is formed by multiplying P by the vector normal to the point. The vector normal to a point is calculated as the average of the vectors normal to the edges to which that point belongs. The normal vector points outward from the body (as viewed from the center of gravity).
8 Correct the length of all springs. To do this, for each spring with an equilibrium distance d0, calculate the difference between d0 and the current length, and if it is greater than D% d0 (initially D = 10) – move the points symmetrically to or from each other so that the distance is equal to D% d0.
9 After calculating the forces in the points 5–7 and integrating the equations of motion with them, and applying the correction to the length of the springs, make a visualization in 2D, black lines representing the springs, minor points as masses, and draw the normal vectors to the points in red.
10 Allow the user to change all the listed parameters: W, H, R, C, f, k, kd, D.
11 Allow pause, restart, etc.

In Figure 33.1, I present a minimal simulation done with the above algorithm and the code generated by Prompt 33.

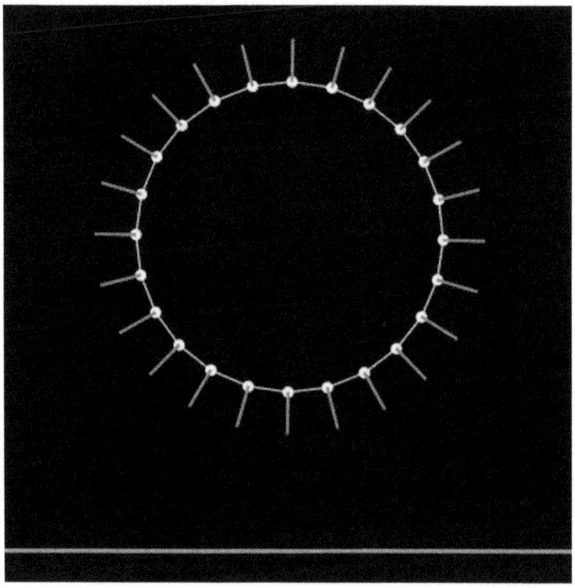

(a)

FIGURE 33.1 Soft-body model of an inflated ball: a) initial moment, b) the ball fell to the surface and deformed, c) the user grabbed one of the points and deformed the ball by moving it upward.

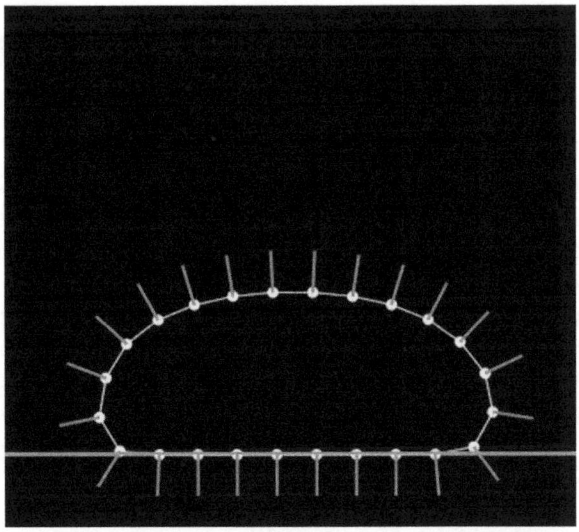

(b)

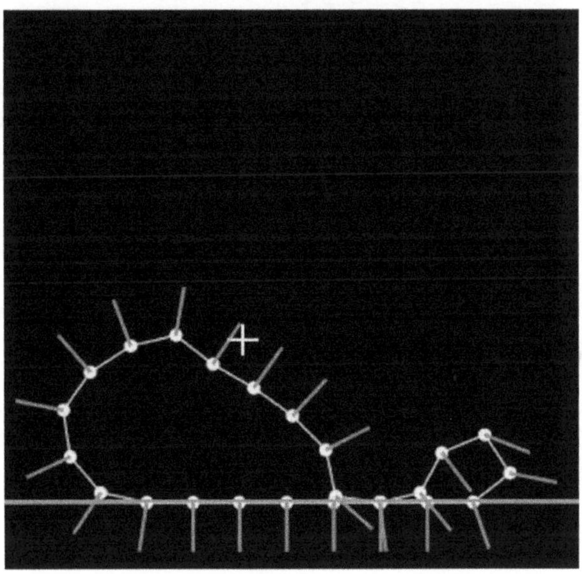

(c)

FIGURE 33.1 (Continued)

Task 1. Generalize the model to three dimensions; generate a sphere object with uniformly distributed points, and a mesh composed of triangles built on this basis. Run the model in a similar situation to the two-dimensional one (including visualization).

Task 2. Add angular forces to the model, which result from additional constraint (angular spring) acting on adjacent edges. These constraints should help the stability of the model.

Task 3. Create a simulation in which a hole appears in the body (by removing one of the walls). The resulting hole should cause the body to move in the opposite direction (a situation similar to a child's game in which you let go of an inflated, open balloon and it moves in random directions). Can you modify the model to account for this movement and changes in direction with the soft-body gas exiting?

34 Ballistic Deposition (Cluster)

Let us consider a model of cluster construction in which particles are deposited by shooting. The cluster will grow starting from the embryo, which is initially placed in the center of the system. At each step, a shot will be taken from a random location on the circle that is larger than the radius of the current cluster. The shot will be taken toward the center of the layout. A radius will be tracked that, when it hits the cluster, will cause the cluster to grow by an additional cell at that location. This is a similar model to the DLA model discussed earlier, with the difference that here we do not have randomness during the execution of the movement of the embedded particle (only during the selection of its starting location). We can add random motion (e.g., with a specific frequency) in this model, which will enrich the model and produce more interesting and DLA-like structures. Let us write such an algorithm taking into account this modification:

1 Work on a W × H grid, with integer values of 0 or 1.
2 Place an embryo in the center of the grid.
3 Randomly select a point on a circle with center in the center of the array and radius W.
4 Trace the radius from the drawn point to the center of the layout (iteratively, move along the unit vector in that direction multiplied by dx/2, where dx is one pixel). Find the cell of value 1 in which the ray strikes first. Fill in the cell from which the ray came by hitting that cell.
5 Return to step 3. Finish if the radius of the resulting cluster reaches W/2.
6 Draw the cluster on a canvas of size bW × bH, black color 1, white color 0.
7 Let the user choose W, H, dx, and b.

The result of the algorithm implemented in the program generated with Prompt 34 is shown in Figure 34.1. Unexpectedly, an interesting cloverleaf pattern emerged, which was created due to the distinct main directions on the network.

Task 1. Run the same model on a triangular grid and check whether the asymmetry associated with the grid directions on which the model operates is still visible in the cluster.

Task 2. Generalize and run (with visualization) the same model in a three-dimensional version.

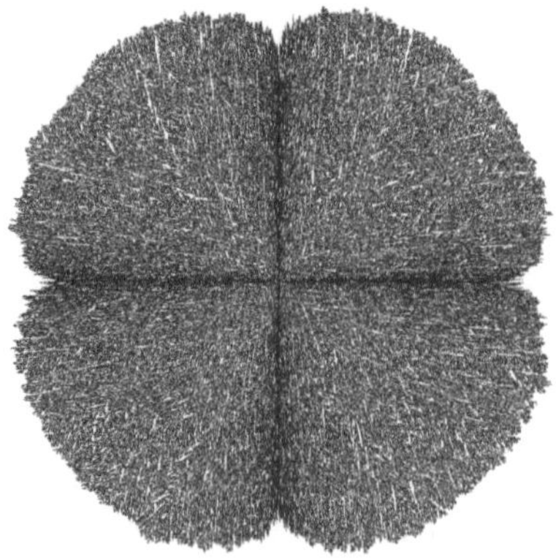

FIGURE 34.1 Cluster growth model with ballistic deposition. You can clearly see the symmetry of the network on which the model was built (x and y directions highlighted).

DOI: 10.1201/9781003734062-35

35 Traffic Jam

Have you ever wondered where traffic jams come from, for example, on a straight highway? Sometimes there is no accident, and spontaneously, in some sections, cars block, and you have to wait until the passage is "uncorked." I put up with it with peace of mind and wait patiently, but I know that many people are very annoyed by this phenomenon. My advantage lies in the fact that I have seen a model of such a traffic jam in action. A straight road, a string of cars, each going at some speed (but they cannot run into each other). Under certain specified conditions, this model will generate fluctuations in the density of cars. Oops. It will simply generate a traffic jam!

Let us assume this will be a one-dimensional model for simplicity. Let it be a discrete array, where values of 0 indicate that there is no vehicle present, the space is free, and it can be occupied by a passing car. This one will be labeled (yes, not otherwise) as a positive number, which represents the speed of the car and the number of cells by which the car wants to move in the next step. For simplicity, we will assume that the highway is infinite; that is, we will loop it in such a way that the last cell is a neighbor of the first one (then you can look at it as a circular highway.

Our program operates according to a straightforward algorithm. In each step, we randomly choose a car and try to move it by the value of its speed. If we cannot do it (e.g., in the bottom place or on the way something is standing), we move it to the car that is standing there (a cell earlier) and set the value of the speed of its neighbor in front (so that in the next step they can drive together). The model we are discussing is an implementation of the Nagel-Schreckenberg type model, about which the interested reader can find more information (I encourage you to consult the references) here (Nagel and Schreckenberg 1992; https://en.wikipedia.org/wiki/Nagel – Schreckenberg_model). Here, however, instead of directly realizing the cellular automaton, we will perform an agent simulation with a random selection of the agent that will perform the shift. However, we expect similar results because the principles are very similar to the original model. So, let us check – for what conditions will traffic jams arise on such a "highway"?

So, to check this, let us write down the algorithm of one-dimensional movement of cars on the highway according to the recipe above:

0 Constants: VMIN = 1, VMAX = 5, DT = 1
1 Create an array of size N and integer values >= 0. Apply periodic boundary conditions.
2 Each cell of the array corresponds to a piece of highway. It contains at most one car (it then has a positive value from VMIN to VMAX indicating the car's speed) or a value of 0 (the space is empty and may be occupied by a car).
3 Arrange the random cars in the array randomly with a density of P = 0.5. Draw the speed of the parked cars.
4 Start the simulation loop:
5 Select a random auto
6 If the car has no other cars in front of it on the way, move it by the distance VMAX*DT. Accelerate it (taking into account the VMAX limitation).
7 If it has other cars in front of it, let it reach the nearest car it has in front of it and reduce its speed by 1 (taking into account the VMIN limitation).
8 Perform points 5–7 in a loop.

The algorithm was implemented using Prompt 35b (for comparison, refer to Prompt 35a, as described in the book's introduction). The results are presented and discussed (captions) in Figures 35.1, 35.2, 35.3, and 35.4.

DOI: 10.1201/9781003734062-36

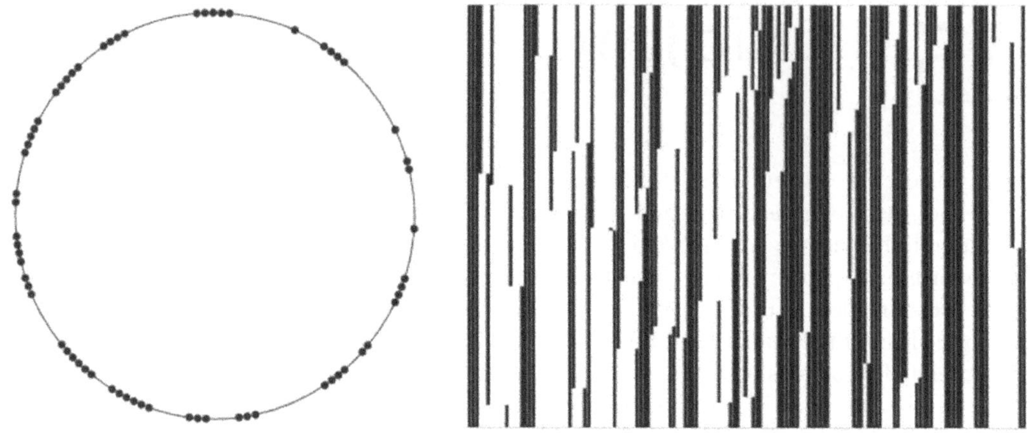

(a) (b)

FIGURE 35.1 Left: highway with cars (circles) and right: density distribution, as a time evolution of the distribution of cars (drawing over time), where each line represents the distribution of cars on the circle over time in such a way that time runs from bottom to top, so the most current line is the one at the very top and it corresponds to the configuration of cars drawn on the circle on the left. The overall density of cars is constant, with P = 0.5 and VMAX = 5.

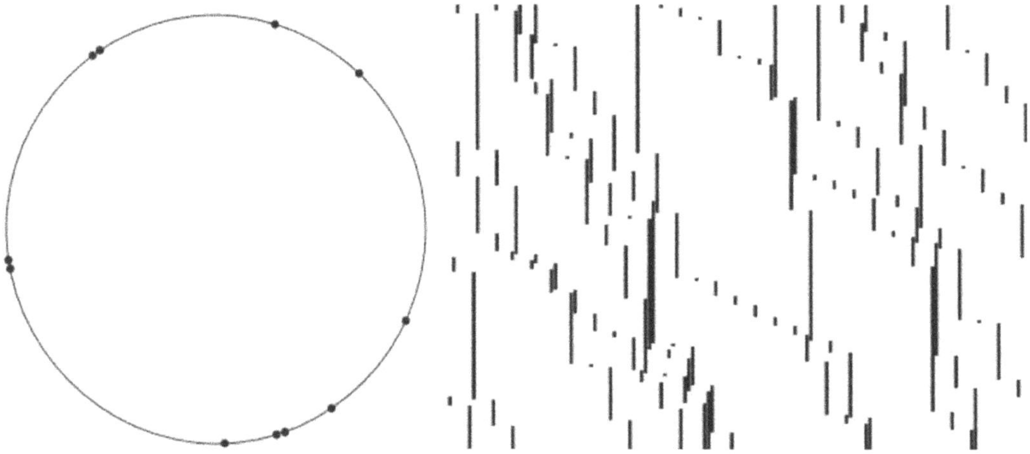

(a) (b)

FIGURE 35.2 The case for P = 0.1 (few cars) with speed VMAX = 5. It can be seen that the low density of cars is not conducive to traffic jams. Traffic is relatively smooth.

Subsequent configurations (line by line) are drawn every fifth time step. This result is for step t = 2345, and as you can see, not much else is happening. We have a group of clusters of cars, but no single giant traffic jam has formed; instead, there are many smaller ones, which make travel very difficult. Cars travel from one jam to another, which is related to their high overall density.

The conclusion of this analysis: if there are really many cars, then the highway is built of one big, clogged traffic jam. However, it is most likely not the density of cars, but the maximum speed that is of primary importance for the formation of traffic jams on the highway under medium-density conditions.

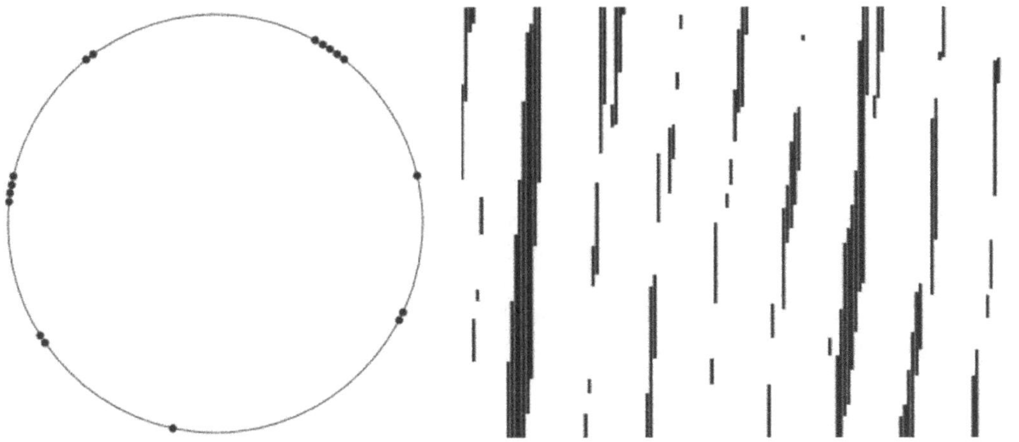

(a) (b)

FIGURE 35.3 Case for P = 0.1 (few cars) with speed VMAX = 15. Time step t = 2024. Traffic is not smooth – cars want to go too fast, and thus traffic jams are created!

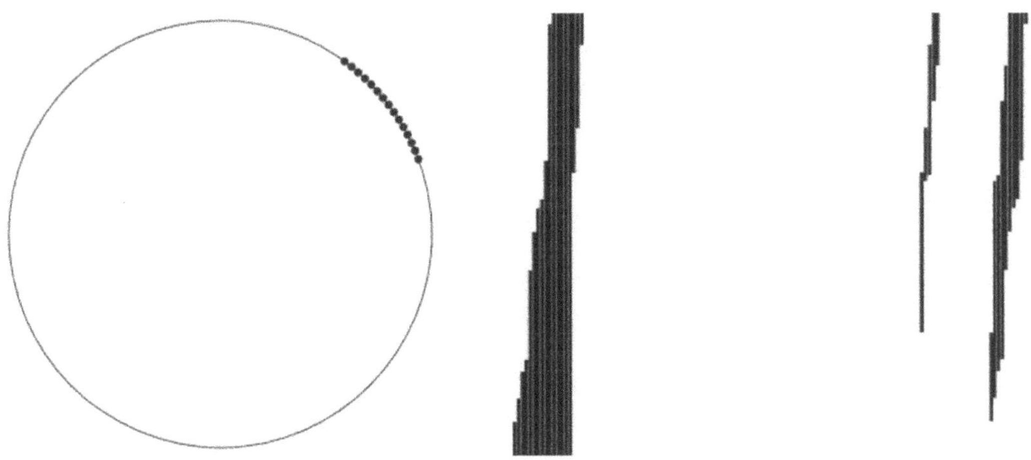

(a) (b)

FIGURE 35.4 P = 0.1, VMAX = N (infinity), clearly a single traffic jam is formed. This traffic jam is stable and will never "spread."

Task 1. Using the model in question, perform a simulation of an intersection with traffic lights and see under what conditions the length of the traffic light cycle can create traffic jams. Is it possible for such a simple simulation to conclude something useful for the practice of designing such intersections?

Task 2. Design a complex layout of roads, intersections, and traffic circles that reflects the plan of your city using a diagram of major roads or a simple map. Run your model on such a layout and try to identify where traffic jams occur.

36 Chromostereopsis

Optical illusions can amaze, especially when we do not employ sophisticated techniques or physical phenomena but rather the properties of our visual organ and brain that cause us to misinterpret what we see. One such illusion is chromostereopsia, in which the right combination of blue and red colors causes the illusion of depth. This is related to the structure of our eyes and how we perceive wavelengths of light. Let us write down a simple algorithm for a technique to create this type of 3D illusion, in which the colors and the small dots drawn with them can give this illusion, just draw them in the right colors and sizes:

1 Create a canvas with a size of L × L, with a black background.
2 In a circle with radius R = L/2 in the center of the layout, randomly draw N1 = 100 circles with radius r = 5px in red color col1 = (255, 0, 0).
3 The whole canvas, excluding the circle with radius R2 = L/2, is placed in the center of the layout; randomly draw N2 = 100 circles with radius r with blue color col2 = (0, 0, 255).
4 Blur the drawn circles gently so that they are not sharp.

The result of a program implementing the algorithm written using Prompt 36 is given in Figure 36.1. We encourage readers to experiment with this – many more interesting shapes can be created using this technique.

Task 1. Generate other figures, more complex, and look for shapes that further emphasize the effect of three-dimensionality not only through color but also through shade and shape of objects and figures.

Task 2. Experiment with colors. Do only red and blue give the illusion of three-dimensionality? Try several different color combinations and find one that works best for you.

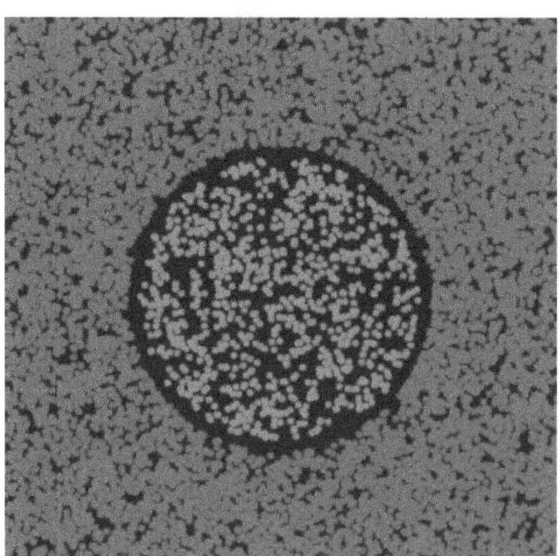

FIGURE 36.1 Chromostereopsia illusion (color figure) generated with given algorithm, where the red (light), center circle is visible as being in front of the blue (dark) background (apparently, some people see it the other way around) (color figure available online).

DOI: 10.1201/9781003734062-37

37 Strange Attractor

A strange attractor is a figure created by iterating the position of a point in two-dimensional space, in which the transformed coordinates of the point converge to a particular closed cycle (attractor), thus drawing patterns and figures. This is a relatively simple procedure, in which we start from an initial position (x0, y0) and iterate through this position in a loop, each time determining a new position. The example we discuss is taken from R. Baumann's book (see Baumann 1989, p. 135, example 22), where we perform iterations:

$$x(n+1) = y(n) + h(x(n)),$$

$$y(n+1) = -x(n) + h(x(n+1)),$$

where the numbers in parentheses at the coordinates denote the time step, n is the time step number. The function h(x) is defined with one parameter a:

$$h(x) = a*x + 2*(1-a)*x*x/(1+x*x),$$

where a (the free parameter) is a number in the range $-1 < a < 1$. The choice of x0, y0, and the parameter a will directly affect the resulting image. The algorithm performing this procedure is simple:

1. let $-1<a<1$.
2. We work on a canvas of W × H (800x600).
3. Start at the center of the layout (x0 = W/2, y0 = H/2).
4. In a loop, perform the transformation N times:
5. x(n+1) = y(n) + h(x(n)), where the numbers in parentheses at the coordinates denote the time step, n is the time step number, and
6. y(n+1) = −x(n) + h(x(n+1))
7. Let the function h(x) be defined as h(x) = a*x+2*(1−a)*x*x/(1+x*x).
8. Visualize successive iteration points.

Figures 37.1 and 37.2 are the results of the algorithm implemented in a program written using Prompt 37. The prompt, in addition to the basic algorithm, generates code that allows you to browse through successive configurations of attractors and parameters (all to find the one from the original book by R. Baumann).

Task 1. Find the parameters for which the attractor resembles the one in Figure 37.1. You could find parameters that better reproduce the figure shown in (Baumann 1989) (page 137 of the Polish edition of the book).

Task 2. Generalize the attractor to one that has an additional parameter b that somehow modifies the iterated function (it can be constant times some factor, an exponent over some word or bracket, etc.). Try to modify the formula in such a way as to get new classes of attractors and patterns generated by the above algorithm.

Task 3. Is an attractor of this type possible in 3D? If so, propose a pattern and carry out iterations and 3D visualization for different parameters.

DOI: 10.1201/9781003734062-38

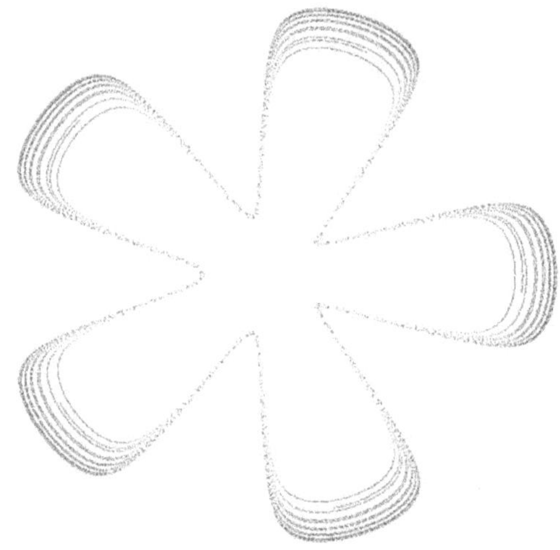

FIGURE 37.1 Example of Henon's attractor (strange attractor), the closest image to the original from (Baumann 1989) that I was able to get. Parameter selection is particularly enjoyable here due to the variety of patterns generated by the chosen iteration equation (see Figure 37.2).

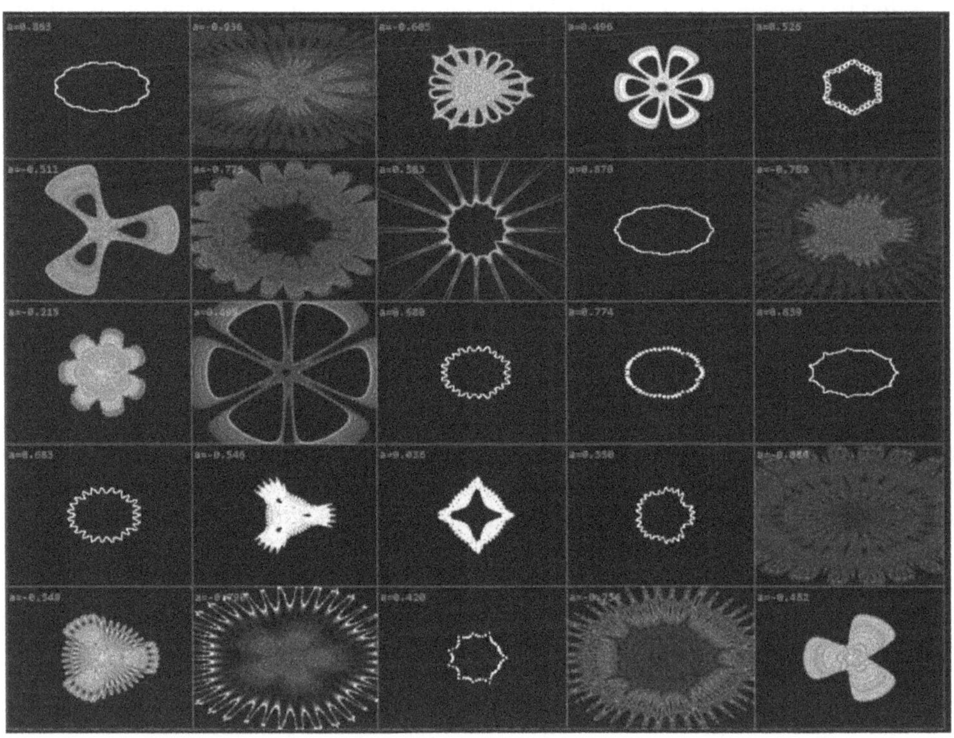

FIGURE 37.2 Henon attractor iteration equations with different parameters "a," for x0, y0 = 10.0.

38 The Lattice Boltzmann Method (LBM)

Computational Fluid Dynamics (CFD) is a field at the intersection of physics, fluid mechanics, and computer science. Within its framework, solvers are created – special programs and algorithms that solve hydrodynamic problems, body flows, flows through channels, droplet behavior, and flows through porous media, among others. In a significant simplification, with many assumptions overlooked, fluid motion can be described by the Navier-Stokes equations. We will not cite them here, not because they are not interesting (we refer interested readers to one of the many books related to fluid mechanics) but because in this chapter, we deal with a model that implicitly solves these equations. This is because it turns out that instead of solving nonlinear differential equations, such as the Navier-Stokes equations, one can solve a simpler problem: the evolution of the probability distribution function of finding particles of a given velocity at a given location in the system. Knowing this function (this is the so-called distribution function) allows us to calculate the macroscopic velocity and pressure fields, which basically means that we have the solution to our problem. The model in question is the Lattice Boltzmann Method (LBM). This is a so-called mesoscopic method, in which we neither track individual molecules nor work directly on the physical fields of velocity or pressure but rather precisely describe the fluid at an intermediate, mesoscopic level, at the level of the velocity distribution function introduced above. This function, depending on the chosen model, can be discretized in various ways. One of these methods is a two-dimensional function f(x, t) described on a square grid with nine allowed velocity directions at each node (the so-called D2Q9 model). The time evolution of this function is described by the Boltzmann transport equation, which can be found in the literature; see, for example, in Polish (Matyka 2021) or English (Succi 2021; Krüger et al. 2017). The LBM model itself is simple to implement, can be written in a very short code, is relatively stable in operation, and is accurate in solving complex boundary conditions. Among other reasons, it has gained tremendous popularity.

Of course, this brief description may have roughly outlined the method itself, but it is not complete and does not even pretend to be so. Its purpose is to interest you, the reader, and perhaps, thanks to it, you will look into its more detailed description in the indicated literature. Let us now write down the algorithm that describes how to perform simulations of flow through a channel with obstructions, using the Boltzmann lattice gas method:

1 Implement the LBM (the Lattice Boltzmann Method) in D2Q9.
2 Perform a simulation of flow through a channel with a circular obstacle in the center.
3 Present the results as a velocity vector field (vectors scaled by a constant), white background, black: vectors, obstacles, particles, and user interface.
4 Periodic grid in x-axis, no-slip conditions on top walls, bottom walls, and obstacles (for no-slip conditions, use the bounce-back technique)
5 Allow modification of gravity, accelerating the flow.
6 Add the ability to change the canvas size and resolution of the LBM mesh in both directions.
7 Use massless particles to visualize the flow trajectory. The particles should initially be arranged randomly and move along the calculated velocity field. If a particle finds itself in a cell where there is an obstacle, or goes outside the system, insert it at a random place on the grid.

DOI: 10.1201/9781003734062-39

The algorithm presented is unique. The language models I want to use for implementation here are taught various algorithms, including the LBM method. So, there is no need for us to go too much into how it works, and we can deal here with the description of the whole issue to be solved. This is a novelty in terms of description in this book. So far, we have described the models and their algorithms in great detail, introducing all, or at least most, of the elements so that the language model implements them at an elementary level by following the written rules or procedures. Here it is different. In the first point of the algorithm, the request to implement the LBM method is written. In points 2–14, the problem is described in detail, including the configuration of the mesh, the presentation of the calculation, the details and parameters of the implementation, the method of fluid acceleration, the collision model on the edges, the boundary conditions, and the size of the obstacles. It appears that we are writing a script to perform a simulation using ready-made software, rather than implementing the LBM algorithm specifically.

On the other hand, if one were to remove everything defined in lines 2–14 from the program created using Prompt 38, there would be little code left, because the LBM algorithm itself is not very extensive. Although this approach is not commonly used in this book, we have decided to retain the fluid flow in this form. This is because if we were to describe the LBM algorithm, we would have to introduce some equations, a description of statistical physics, and so on. All this would obscure the simulation itself and the physical phenomenon under study. Such a description of the algorithm also makes it easy to add elements to the program related to, for example, the calculation of various quantities on the velocity fields obtained. Thus, we added calculations of the tortuosity and porosity of the system, which required an additional sentence to explain what these calculations entail. Figure 38.1 is the result of a program written with a single Prompt 38.

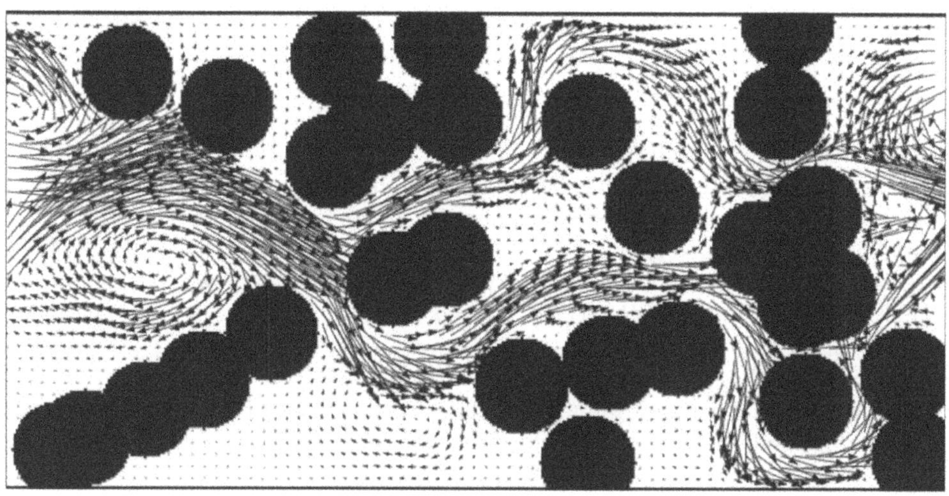

FIGURE 38.1 Fluid flow through porous media built of overlapping circles made of randomly distributed obstacles in the x-periodic and y-no slip geometry (pipe). Computational grid 400 × 200, relaxation time $\tau = 0.61$, gravity $g = (0.0005,0)$. The system has a porosity of 0.6017 and a tortuosity approximately $T \approx 1.258$ (you may compare the value, i.e., with (Matyka et al. 2008)). Scaling the vectors allowed us to visualize several vortices, with the most visible formed in the top left, middle-left, and middle-bottom parts of the system.

Task 1. Perform the exact simulation in 3D with visualization of particles flowing in the fluid region.

Task 2. After calculating the steady state, draw current lines showing the trajectory of particles flowing through the entire system.

Task 3. Add procedures to the program that calculate the properties of the system as a porous sample. At each time step, recalculate the steady state condition as maximum change of velocity field between two timesteps at times t and t+50 computed consecutively in each cell in the flow simulation. Display it as single number, similarly to other data. Add a button to randomize the configuration of obstacles and slider to vary overall porosity from 0.35 to 1.0 for configurations generated using that button. Add slider to control delta time used for massless particle time integration in animation (from 0.1 to 5.0 with step 0.1). Add plot of tortuosity in time that is plotted on the website as separate canvas to see how it evolves in time. Please add a separate canvas with the plot of tortuosity vs. porosity in which you add points measured from the main simulation when the user hit "Add Result" button. Please also add "Clear" button that clear the plot of tortuosity vs. porosity.

39 Agent-Based Epidemic Spreading

When the COVID-19 pandemic began, models – often based on simple agent-based simulations – that predicted the spread of disease in a community became immensely popular (https://www.washingtonpost.com/graphics/2020/world/corona-simulator/). Such models can be built at different levels of complexity, with numerous parameters, including social distance and mask usage. Such a model, in its basic version, is straightforward and boils down to tracking the movement of agents, represented, for example, by circles, which move through an area, experience collisions (contact) with each other, and in the course of such collisions may exchange diseases. In our model, we will assume that initially a certain fraction of agents will carry a disease. In the course of contact, they may (but not necessarily) transmit it to each other, which will also be related to probability. An additional mechanism will be the possibility of the death of agents who are already sick, and here again, probability will be helpful.

It is worth noting that models of this type can also be described by differential equations, writing how the fraction of healthy and susceptible (susceptible), sick (infected), and resistant or deceased (removed) agents changes, and they carry the name SIR models from the first letters of these English terms (https://en.wikipedia.org/wiki/Compartmental_models_(epidemiology)). You can show that for the same parameters, both approaches produce the same results, with the observation that agent models give much more freedom and flexibility to add new mechanisms. So let us write down an algorithm that implements this type of model. The description of the algorithm will be at a level where we are not so interested in the implementation details of collisions between agents but more in how and with what parameters contagions can propagate.

Let us prepare algorithm for an agent-based epidemic spreading:

1 Simulate N agents (circles of radius r) moving in a room of size L × L, with rigid boundary conditions.
2 The circles interact with each other in a spring-like manner (collisions). Use a model with a virtual spring so that agents can gently step on each other without causing harm.
2 Agents move in a uniform rectilinear motion, only changing direction of motion when colliding with other agents.
3 Initially, all agents have a status of "Healthy" represented as a green color in the simulation. N2 agents selected randomly get the status "Sick."
4 When a healthy agent comes into contact with a sick agent, the sickness passes with probability p from the sick agent to the healthy agent.
5 During the simulation, each sick person dies with probability p2.
6 The simulation stops when all agents are sick or have died.
7 During the simulation, draw on a single graph:

Z(t) – healthy since time in green, C(t) – sick since time in red, U(t) – dead since time in black.

Figures 39.1 and 39.2 are the results of a program simulating agent-based spreading of disease using the above algorithm and written using Prompt 39.

Task 1. Add to the program the analysis of the problem with the differential model SIR in which dS/dt = –B/N * (I*S), dI/dt = B/N*(I*S)-g*I, dR/dt = g*I, where B, g are parameters, S are healthy, I infected, R immune or dead, N is the number of agents, and t is

DOI: 10.1201/9781003734062-40

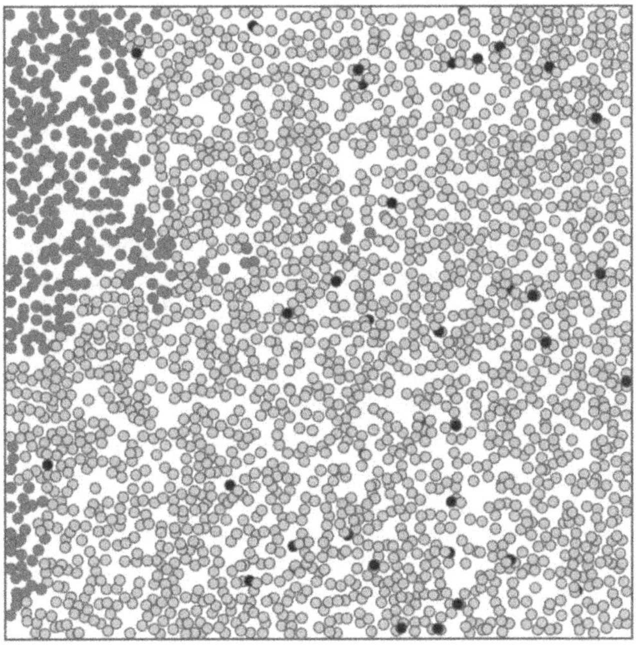

FIGURE 39.1 A selected moment from a simulation during an epidemic spreading through a population of 2500 agents performed according to the algorithm introduced above. Most of the population is already infected (red/light gray cluster), some are healthy (green/dark gray), and some individual agents have died (black) (color figure available online).

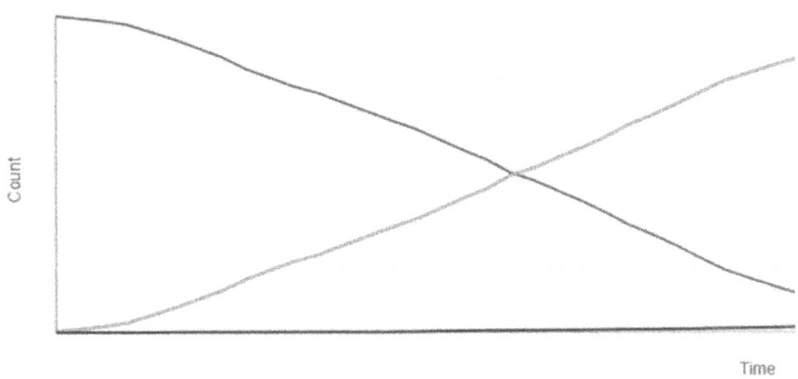

FIGURE 39.2 A graph showing the evolution over time of the system from Figure 39.1. You can see how healthy individuals turn into diseased ones. There is also a slow extinction, which, however, does not dominate the system (color figure available online).

time. At the end of the simulation, draw one graph where the results of the agent model are drawn together with the results of the SIR model with appropriately chosen initial parameters.

Task 2. Study the behavior of the system and the dynamics of the epidemic in the case in which agents can make quick, random jumps to any place in the system. To achieve this, add new parameters to the layout, such as the frequency of these jumps and their duration. See if such "trips" have a significant impact on the epidemic.

40 L-System Tree

L-Systems are a special class of algorithms and a way of writing figures composed of iterative structures (https://en.wikipedia.org/wiki/L-system). It is an exciting technique in which we start with a particular state, a word (or string) that serves as the trunk of our figure, and in successive iterations, apply repetitive, clearly defined operations to its successive letters. These operations generally make our figure grow (each iteration adds new elements to the word). If our initial state were "X" and the rule were "X->XX," then in subsequent iterations, the letters would multiply without resistance; that is, successive iterations would look like this: "X," "XX," "XXXX," "XXXXXXXX," etc. This is a trivial example, but it effectively illustrates the idea.

The letters in our growing word (state) are represented as a string and later drawn. This is done in such a way that each letter represents a distinct graphic operation. Usually, the so-called turtle graphics are used here, that is, simple operations like "move," "pick up the pen," "go back to the beginning," etc. In this way, you can create amazing patterns, fractals, Cantor sets, curves, etc. We will consider some standard examples of building a simple tree, which is written in the form of an L-System, and has the following initial state:

$$X$$

and rules:

$$(X \rightarrow F+[[X]-X]-F[-FX]+X), (F \rightarrow FF).$$

What will the evolution of the state look like in subsequent iterations? Obviously, the first iteration is the initial state, and subsequent iterations are a direct consequence of the aforementioned rules:

1: X
2: F+[[X]-X]-F[-FX]+X
3: F+[[F+[[X]-X]-F[-FX]+X]-F+[[X]-X]-F[-FX]+X]-F[-FF+[[X]-X]-F[-FX]+X]+F+[[X]-X]-F[-FX]+X

. . .

We will not elaborate on the rules further here, as the algorithm typically handles that. To complete the operation of the program, we still need to define how to draw such a state. Well, here we use the technique (as we mentioned earlier) of the turtle, which, stepping, draws the graphic and so:

F means moving forward (for example, by 1 pixel) while drawing a line
+ is a rotation to the left by 25 degrees
– is a rotation to the right by 25 degrees
[is to save the current state (position and angle of the turtle) on the stack
] is to remove the state of the turtle from the stack

X we leave – it does nothing to the turtle and is used for the evolution of the L-System. We will now write the algorithm summarizing all the above rules:

1 We work on the canvas W × H
2 Let state = "-X"

DOI: 10.1201/9781003734062-41

3 Create a map of rules for swapping successive state elements, for example: (X → F+[[X]-X]-F[-FX]+X), (F → FF)
4 Allow a single iteration on the state: the loop goes through all the elements and executes:
   (a) "[" saving the current state (position and angle) on the stack
   (b) "]" removing the state from the stack
   (c) "F" move forward by FWD pixels and draw a line (e.g., FWD = 1)
   (d) "+" rotate by ANG degrees to the left
   (e) "-" rotation by ANG degrees to the right

Figure 40.1 shows the tree generated using the above algorithm and the set of L-system rules encoded within it. The program that drew it was generated using Prompt 42.

Task 1. Write down another L-System and try to use the Koch Curve, Dragon Curve, and Sierpiński Triangle. A good source of inspiration can be the entry on Wikipedia, where the rules ready to be introduced into the program are given (https://en.wikipedia.org/wiki/L-system).

Task 2. Program the set of rules in the 3D case and add the ability to visualize the L-System in 3D, as well as draw some sample trees.

Task 3. Set the L-Systems in motion by making sinusoidal changes in their structure, for example, by applying (depending on the number of a given segment) changes to the angles used when a turtle walks.

FIGURE 40.1 Tree made using L-Systems and the algorithm described earlier for a set of conditions without initial trunk rotation (simply "X") and eight iterations with rotation 30 degrees, forward distance 1.

41 Dragon Curve

A dragon curve is a fractal figure that can be created either by recursively bending a segment (strip) at a suitable angle, which can be implemented, for example, using the L-System from the chapter on generating trees, or by iterating functions (the so-called Iterative function system IFS). In this chapter, we will learn about the latter approach and calculate the iterations of a system of two functions that determine points within a specific figure, known as the title dragon. Algorithms of the IFS type are elementary and reduce to calculating the successive positions of points that can be drawn during iteration. In the case of the curve in this chapter, there are two such functions, and we will choose them randomly. We start from the point (0, 0) and iterate over two functions:

$$f1(x, y) = (1/sqrt(2)) * R(45) * (x, y),$$

$$f2(x, y) = (1/sqrt(2)) * R(135) * (x, y) + (1, 0),$$

where R is the rotation matrix in two dimensions, and the number in parentheses specifies by what angle we rotate. Let us write down the above information in the form of an iterative algorithm:

1 We work on the canvas W × H.
2 Start with (x, y) = (0, 0).
3 Iterate the following functions (choose f1 or f2 randomly): f1(x, y) = (1/sqrt(2)) *R(45) * (x, y), f2(x, y) = (1/sqrt(2)) * R(135) * (x, y) + (1, 0), where R(phi), is a 2 × 2 matrix of rotation by phi degrees.
4 Draw successive iterations as points on the 2D plane.

The result of executing this algorithm with a program generated using Prompt 41 is shown in Figure 41.1.

FIGURE 41.1 The dragon curve that resulted from iterating a function of two variables (successive values of the function are the points drawn in this figure.

DOI: 10.1201/9781003734062-42

Task 1. Experiment with the values of the phi angles chosen when iterating to obtain other interesting patterns from the iteration. Here, we obtained a seahorse for (22, 222) and a slightly inverted fern for (–5, 33). You can also (although this will already require changes to the algorithm and may be difficult to control) play with the f1 and f2 functions themselves.

Task 2. Generate an animated Dragons in which successive frames correspond to a Dragon's Curve for changing angles, parametrizing of matrices R in both functions.

42 Granular Matter with Friction

If we were to drop grain after grain, then, after a certain length of time, a cone of a particular angle will form, which is characteristic of the material we are pouring. Granular materials, such as sand or other loose materials composed of small, relatively stiff grains, exhibit several interesting behaviors. Many of these are a direct consequence of particle interactions, and one of the most interesting is friction between particles. Such friction can be dynamic or static, and it is the latter that is the "hero" of this chapter. Owing to the existence of static friction, the granular material can form a characteristic and stable angle of rest in the heap mentioned previously. This static friction makes it possible to play a game of Mikado (see Figure 42.1). Here, the static friction between the beads causes the arrangement to be stable, and chains of forces are formed to maintain a stable configuration.

In 1993, J. Lee and H. Herrmann proposed an interesting frictional granular matter model to simulate the motion of rigid disks (particles) in a field of gravity interacting at the moment of contact by repulsive and friction forces (Lee and Herrmann 1993; Ristow and Herrmann 1994). The proposed model incorporates several forces that, in total, accumulate to a single force vector F(j->i) acting on particle and originating from the contact with particle j. This force is generally composed of a normal and tangential part: F(j->i) = Fn *n* + *Fs* s. We will omit the exact description and expressions for the normal force (these are written in the following algorithm) and focus on describing the phenomena of static friction. The force Fs appearing in this equation also consists of two components:

$$Fs = -\gamma s*me*(v*s) - sign(\Delta S)*min(ks*\Delta S, mu*Fn),$$

where the first term is the energy dissipation associated with the relative motion of the bodies (v is the difference in their velocities). More interesting is the second expression, where ΔS represents the distance one body has traveled relative to the other during contact, and the force ks*ΔS acts opposite to the relative motion of the bodies. In the formula given earlier, we use Coulomb's criterion to constrain the force of the starting friction, ks*ΔS, by considering the maximum force that this friction can induce, given the current normal force acting on the body (mu*Fn). The model amounts to the appearance of a spring (hence the force ks*ΔS) at the point of contact, which elongates as long as the bodies remain in contact and is removed when they separate. The spring also ceases to act when its force exceeds a certain level, which is set by the normal force. In practice, in a simulation, the point of first contact must be remembered so that the virtual spring remembers its elongation.

Once the forces are calculated and gravity is added, the equations of motion of the particles can be solved using one of the known methods (we will use Euler's method or Euler's implicit method). It turns out that both work pretty well. Therefore, we write an algorithm that completely describes the simulation in this model as follows:

1 The pellets have a diameter of d0 = 1mm. The physical layout is 10 × 5cm in size.
2 Use the 5th order corrector predictor method, Euler method, implicit Euler, RK2, and RK4 to integrate the equations of motion of the discs (let the user choose the integration method).
3 The granule consists of circular particles.
4 Particles interact when they are in contact with each other.
5 For two particles i and j, their positions Ri and Rj are vectors indicating their centers.
6 Let the vector r = Ri – Rj
7 The vector n will be the unit vector parallel to the line connecting the centers of the two molecules; that is, n = r/|r|, where |r| is the length of the vector r.

DOI: 10.1201/9781003734062-43

FIGURE 42.1 Mikado toy during play – some of the beads have already been removed and what remains are mostly those that hold the pushing spring through static friction.

8 The vector s, perpendicular to n, is obtained by rotating the vector n by an angle of 90 degrees in the clockwise direction.
9 Velocity v = Vi – Vj, where Vi is the velocity of particle i and Vj is the velocity of particle j
10 The radius of particle i is ai; the radius of particle j is aj.
11 Let us define the force acting on particle i and exerted by particle j in contact with it; i.e.,

F(j->i) = Fn * n + Fs * s,

where the normal force Fn is:

Fn = kn * (ai + aj – r) ^ (3/2) – γn * me (v * n),

where kn is the elasticity constant of the material. The sign "^" denotes a power. The first term is the Hertz model. The second term is the velocity-dependent friction term, which is introduced to dissipate energy when bodies collide. ɣn is the constant controlling this dissipation, and me is the effective mass, me = mi*mj/(m+mj).

The tangential force Fs in the force equation is:

Fs = −ɣs*me*(v*s) – sign(ΔS)*min(ks*ΔS, mu*Fn),

where the first term is also dissipation (here associated with motion along the s vector). The second term is static friction. When two bodies start contacting, we insert a virtual spring between them, which extends in the shear direction. For a total displacement of ΔS during contact, there is therefore a force ks*ΔS, which acts opposite to the movement of the bodies. The maximum value of this force is given by the Coulomb criterion of static friction mu*Fn. When two bodies are not in contact – we remove the spring (the static friction force ceases to act). In order for this model to work correctly, it is important to remember the total relative displacement of the bodies in contact in the shear direction (and not just the instantaneous displacement between two time steps).

12 When a body is in contact with a wall, the interaction is the same as with other particles, considering aj = 0 and mj = infty (infinity). The whole simulation takes place in a gravitational field, so there is an additional force -mi*g acting on each particle. The total force acting on a particle is therefore the vector sum of the forces acting from contact with other particles, contact with the wall, and gravity.
13 Visualization: white background, white circles with a 1px wide black border to represent static particles and filled circles in black for dynamic particles, no additional embellishments.
14 Initial values: g = −10m/s^2, mu = 0.2, kn = 10^6, ks = 10^4, ɣn = 5*10^2, ɣs = ɣn/100, and for walls kn = 2*10^6.

In Figure 42.2, we present the results of a model created based on the above algorithm.

Task 1. Please use the model to create the Mikado game shown in the first picture. The game mechanics are based on static friction. Allow the users to take turns removing beads from the game board. Please add the ability to draw tension between beads.

Task 2. Please add to the above simulation the ability to visualize a tree showing the stresses in the system (thicker joints correspond to higher forces).

Task 3. (complex) Please generalize the static friction model (and the entire simulation, including initialization, integration of the equations of motion, and visualization) to the 3D case.

Task 4. (complex) Analyze density waves under a hopper outlet and try to find the 1/f noise predicted in (Ristow and Herrmann 1994).

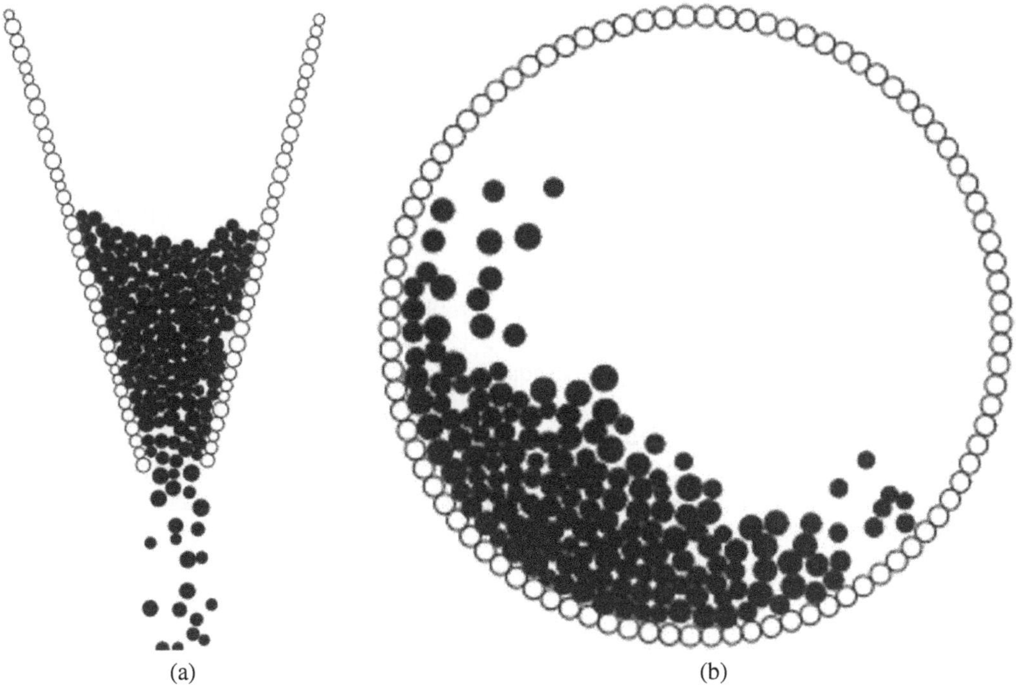

FIGURE 42.2 a) flow out of granular matter with friction from a container, b) granules in a rotating cylinder (program written using Prompt 42).

43 Standard Map

Turbulence has fascinated (and continues to fascinate) humans for centuries. This phenomenon is one of the less-explored classical phenomena, where the combination of chaotic movements of particles and vortices cannot be easily described mathematically. One of the millennium problems is that of fluid motion, which concerns the proof of the existence of solutions to the flow equations (https://en.wikipedia.org/wiki/Millennium_Prize_Problems). Turbulence cannot be solved analytically, and theoretical and numerical models are subject to errors, which, combined with the aforementioned chaos, can have catastrophic consequences.

We will not deal with turbulence directly, but we find that the chaotic and disordered motion of vortex structures, shown as vortex function contours, is very similar to the contours of the chaotic maps generated by the iteration model of two simple functions. The model we discuss is called the Chirikov standard map (https://en.wikipedia.org/wiki/Standard_map) in the literature and is generated by iterating two simple functions:

$$p(n+1) = p(n) + K * \sin(t(n))$$

$$t(n+1) = t(n) + p(n+1).$$

If, for a given parameter K, we draw successive points (p, t) on a periodic board of dimension 2π, we obtain amazing patterns whose shape depends only on this parameter. Now, let us write an algorithm that describes the exact scheme of the standard map generation procedure as follows:

1) Start from point (1, 1) and K = 0.971635.
2) Iterate a pair of equations:

$$p(n+1) = p(n) + K * \sin(t(n))$$

$$t(n+1) = t(n) + p(n+1),$$

   where n is the iteration number. Start from the middle.

   After each iteration, take modulo $2*\pi$ of both p(n+1) and t(n+1).
3) Draw p(n+1), t(n+1) on an 800 × 800 canvas.

In Figure 43.1, we show a map generated using the above algorithm for the single parameter K indicated there.

Task 1. Investigate maps for other K-parameters.

Task 2. Make the parameter K time-dependent and study the evolution of the map over time (animations).

Task 3. Search for other sets of iterative functions that generate interesting maps.

Task 4. (complex): Perform an inverse analysis of a standard map and attempt to obtain the vector field that leads to the map. From such data, fluid flow and particle transport can be generated and investigated, for example, the turbulent properties of a hypothetical fluid flow.

Task 5. (complex) Make it 3D by iterating three functions in 3D space.

DOI: 10.1201/9781003734062-44

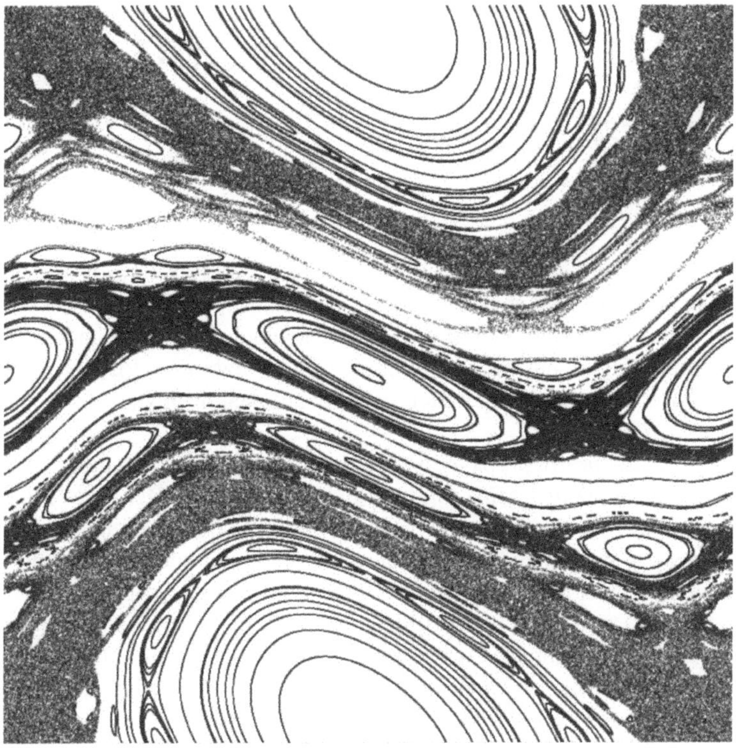

FIGURE 43.1 Chirikov standard map for K = 0.971635 (program written using Prompt 43).

44 Excitation Model and Spiral Waves

In nature, many objects have spiral structures. They have different origins and occur at extreme scales. Examples include the snail shell and the rather distant spiral galaxies.

Rather than modeling a system of billions of particles interacting gravitationally (which would perhaps create a spiral galaxy, but we would have to wait a long time for that), we will attempt to generate a spiral wave in a simple excitation cellular automaton model that is used to track fire propagation.

Let us consider a model of fire propagation in a forest that has several states. A tree in the forest may burn for some time, and during this time, the fire may spread. After a tree has burned, another tree may grow in its place, but this may only happen after some time, which is a characteristic of the model. Such a model can be realized as a cellular automaton in which the fire spreads on a square grid, and each node is subject to the same rules. To do this, two arrays are used to store the state of the system (e.g., A and B), and a given algorithm is executed on one of them, performing operations with a destination in the other (the idea is that local changes do not cause disturbances to neighbors during one time step). After all operations are performed, the arrays are then swapped.

The model allows wind to be added by changing the probability of fire spread and adding obstacles to prevent the fire from spreading toward specific sites.

The model was presented to me during the 8th Statistical Physics Summer School in Chęciny, Poland, where I attended an excellent practical workshop on excitable media. The task to be performed there was a program according to a given algorithm and an attempt to obtain the effect of spontaneous formation of spiral waves. Interestingly, the desired effect could only be obtained by introducing a delay, specified by a certain number of time steps, at which the burned sites of the model became neither tree nor fire (resting state).

Therefore, let us write the complete algorithm of a multi-state excitation model with delay, for a forest fire, as follows:

1. L × L square grid.
2. At each node of the grid, place a tree with probability p1.
3. Use the cellular automaton technique. Work on two copies of the array. Perform the following rules first for all the cells in array A, storing the result in array B, and then copy the whole array B to array A.
4. In a loop:
   - On each cell without a tree and without a wall and not in a refractive state, a tree can grow with probability p2.
   - Any cell with a tree can ignite from an edge cell where there is a fire with probability depending on the direction of the ignited neighbor (neighborhood of 8 directions – 4 main directions and a diagonal).
   - Each tree can ignite itself with probability p4.
   - The tree burns for P simulation steps.
   - After burning, the field is in a state of refraction for REF steps in a state where nothing can grow there, and only then does it turn into an empty field where something can grow.

The results of the model are presented in Figure 44.1.

DOI: 10.1201/9781003734062-45

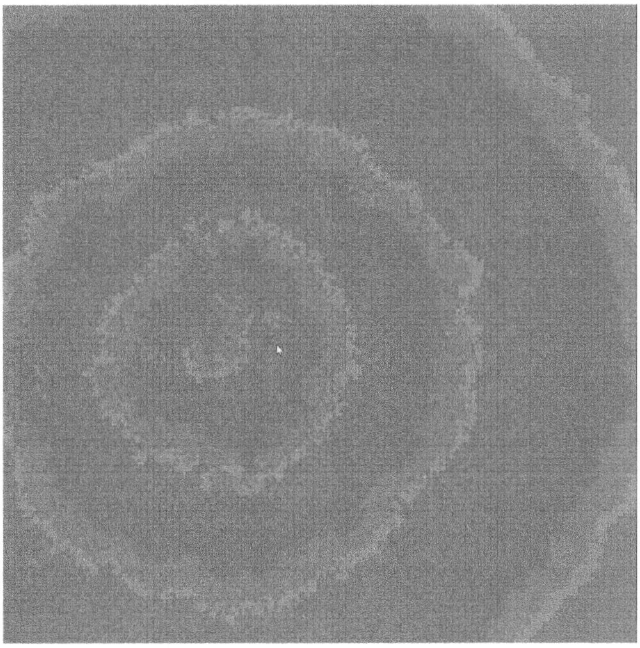

FIGURE 44.1 Spiral waves arising spontaneously in a fire simulation with delay (program written using Prompt 44).

Task 1. Perform a 3D simulation in which a forest fire propagates spatially in a layer with thickness h.

Task 2. (complex) Visualize a spiral wave in 3D using the solution from the previous task.

Task 3. (complex) Write a program that randomly samples the initial conditions and simulation parameters to search for a spiral wave in the solutions (you will need a criterion for a spiral wave based on the analysis of the current state of the simulation). What set of initial values on the grid will yield a stable spiral wave?

Task 4. Is it possible to obtain a spiral wave without the delay mechanism described herein?

Task 5. Define the model and examine its behavior on a triangular and hexagonal grid.

45 Paper, Stone, Scissors

Everyone is familiar with the simple game of paper, stone, and scissors. Paper beats stone. Stone beats scissors. Scissors beat paper. A similar scheme can be applied to a space filled with such objects. It turns out that, unexpectedly, this simple game can be the basis for a model of a complex Belousov-Zhabotinsky-type chemical reaction if the game considers geometry and neighborhood.

If there are more than two scissors in the surroundings of, for example, paper, they will defeat the paper and new scissors will appear in its place. The model is simple because it boils down to counting what is in the surroundings of each point. This can be solved using a simple cellular automaton. Here, we work with an array containing integral values corresponding to successive types of objects. In our case, it will be a three-state automaton, where for successive objects (paper, stone, and scissors), the numerical values 1, 2, and 3 will be entered. Thus, knowing which object defeats which, the algorithm will consist of simultaneously counting the sums of objects of other types in the neighbors, checking whether the number of the object that "defeats" us is greater than some threshold (the standard threshold is 2), and possibly changing the current value. An interesting feature of this model is that, unlike the forest fire model, spiral waves are always created and are stable. Therefore, let us write down the algorithm of the cellular automaton of paper, stones, and scissors:

1 Simulate a cellular automaton with two arrays of W × H labeled as (A, B), where we work on array A; perform on all elements the algorithm writing the results to B; and then make a copy of B->A.
2 For each node, write one of three random values: type 1 = 1, type 2 = 2, or type 3 = 3.
3 In the simulation, consider that type1 beats type 2, type 2 beats type3, and type 3 beats type 1.
3 Simulation loop:
4 From the environment of the current node (periodic conditions), with type TA, count how many cells of other types (TB and TC) are in total, the numbers are NA and NB:
   4a) If TB beats TA and NB > THR -> set the new value in the cell to TB
   4b) If TC beats TA and NC > THR -> set new value in cell to TC
   If none of the conditions are met, leave the current value in the cell.
5 Return to 4.

The spiral waves (simulation frame) showing the results of the above model are shown in Figure 45.1.

Task 1. Generalize the model to N agents, where cyclically the first overcomes the second, the second overcomes the third, . . ., and finally the Nth overcomes the first. Simulate large N (what is the constraint?).

Task 2. Run the same model in three-dimensional space with visualization and determine whether spiral waves are produced for the three reacting factors.

DOI: 10.1201/9781003734062-46

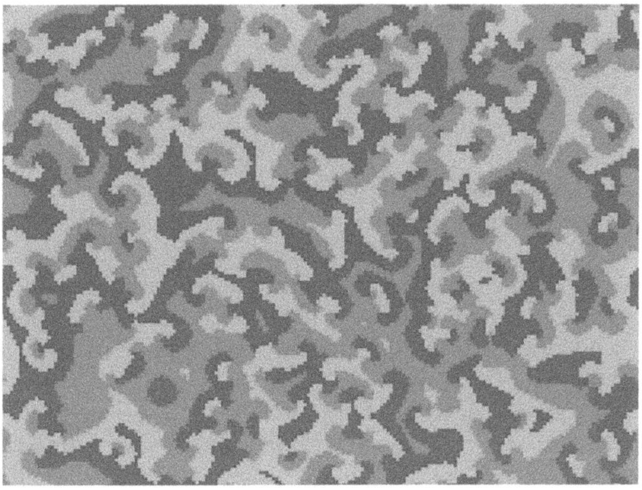

FIGURE 45.1 Periodic spiral waves in the paper, stone, scissors model. Successive colors correspond to different objects (program written using Prompt 45) (color figure available online).

46 Classical Billiards

Let us consider the model of a classical material point, located in a container, that moves in uniform rectilinear motion without external forces (i.e., no gravity) and takes into account reflections from the container's edges. The study of even such a simple system can be interesting if we consider more than one perspective. There is a branch of research and systems in physics known as dynamic billiards (https://en.wikipedia.org/wiki/Dynamical_billiards). There is a quantum version of this model. However, we will consider its classical version here, in which material points move according to the equations of motion of a classical particle.

Simulation of such a system involves tracking points, often distributed with a delicate deviation in a random fashion. If the initial velocity of the points was the same, they may follow a similar trajectory. However, due to reflections and spatial variations in the curvature of the object from which the points bounce, the points can scatter, marking significantly different trajectories. Also mentioned in this context is the Lyapunov exponent, which can be determined for a pair of points whose trajectories diverge over time.

So, let us write down an algorithm in which we describe the simulation of the classical billiards version for multiple (initially randomly distributed) particles bouncing off the edges of figures of different shapes:

1 We are working on a 1280 × 720 canvas.
2 On the board, draw with a thick line a large figure: oval, ellipse, circle, square, rectangle, square with a full circle in the center (circle with radius rk) or rectangle with rounded corners with radius of curvature r1.
3 At the point (x, y), place N points with a random variation of +/– (dx, dy).
4 Give all points the same initial velocity equal to a vector with coordinates (vx, vy).
5 The points move in a uniform rectilinear motion.
6 The points bounce off the edge of the figure drawn in the center of the system.
7 Moving points leave a trace (do not automatically clean).

Figures 46.1 and 46.2 are the results of the program generated according to the algorithm described above, using Prompt 46.

Two points released at the same speed from different (but close) starting points, moving in a square box, follow a pair, and their trajectories do not diverge. Only by placing the circle in the center did the system behave chaotically, and the trajectories diverged quickly. The Lyapunov coefficient describing the speed of divergence of the trajectories is non-zero and is (calculated by the program) $\lambda = 0.0081$.

Figure 46.2 illustrates the significant difference in the model's behavior depending on the shape of the container. In an ellipse, the points are always compactly grouped. In contrast, in a rectangle with rounded corners, the system quickly begins to exhibit chaotic behavior, and the points follow significantly different trajectories. It is interesting to note that in the case of the ellipse, places have formed where the points will never be reached.

Task 1. Add a magnetic field (perpendicular to the 2D surface) to the simulation, and assume that the particle has some non-zero charge. Check how a single molecule, as well as a whole group of them, behaves now.

DOI: 10.1201/9781003734062-47

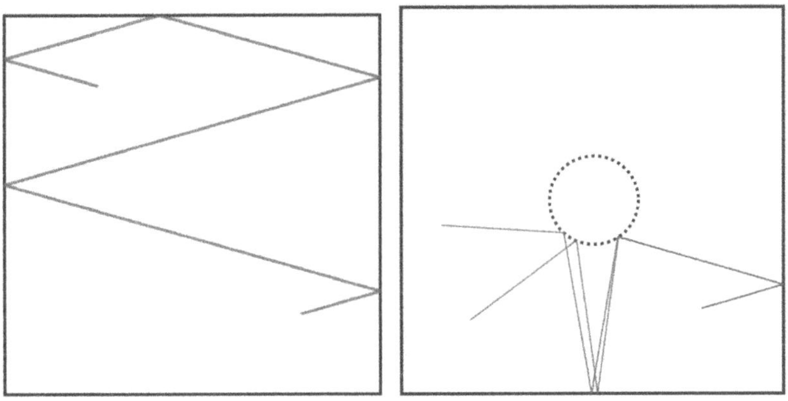

(a) (b)

FIGURE 46.1 A classic billiard of two points in a container without an insert (left side) and with an insert (circle, right side).

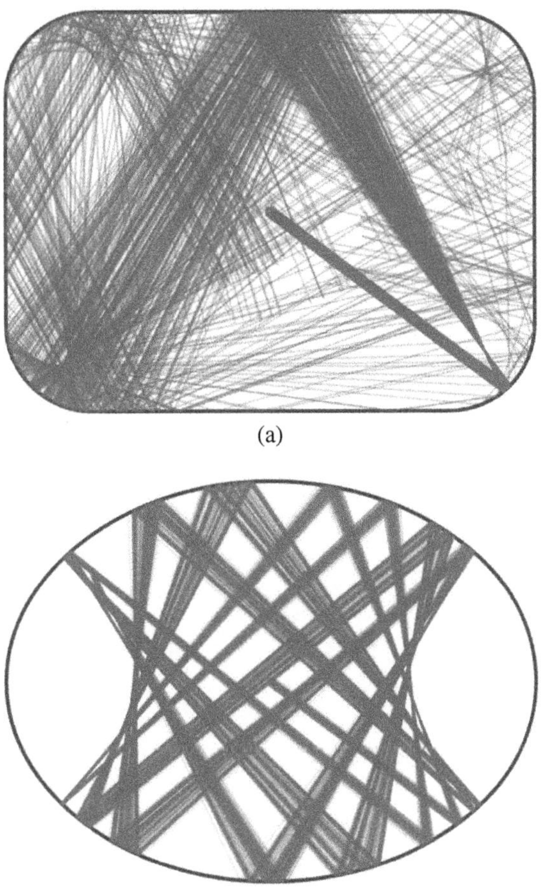

(a)

(b)

FIGURE 46.2 Comparison of simulations of a classic billiard of N = 100 points, initially placed in the center of the layout with a random offset of +/– 6px and with an initial velocity of (2, 1.5). On the left, the behavior in a rectangle with a radius of curvature of the corners equal to r1 = 130px. On the right, the behavior of points in an ellipse of the same width and height.

Task 2. Make a 3D simulation of a billiard and see if you can get and visualize the chaotic propagation of particles.

Task 3. Construct a "Tokarsky room" type geometry and see if you can find obashars that cannot be illuminated by a single light source (see Tokarsky 1995) by trying from several different locations.

47 Planetary Systems

Why doesn't the solar system disintegrate when giant planets, interacting with each other by gravity, move at relatively small distances from each other? What keeps our system from flying away? It is not hard to guess that this something is the Sun, dominating the center of the solar system with a mass hundreds of thousands of times that of Earth. A system of gravitationally interacting planets clustered around a single, larger, and dominant star is stable because of the relatively small interactions between the planets, which do not significantly affect each other but rather "feel" the Sun's gravitational pull. In the case of the three-body, unstable system we modeled earlier, it was different – the simulated bodies had similar masses.

To model a planetary system, you need to do a numerical simulation, similar to the simulation we did for the three-body system, where bodies distributed in space interact with each other according to the law of universal gravity (force inversely proportional to the square of the distance between the bodies; see the chapter on three bodies). In our model, we will apply the law in a version with an interaction smoothing factor. This is a technique often used in modeling, in which instead of using the inverse of the distance between bodies (1/r^2) in the denominator, 1/(r^2 + eps) is used, where eps is a relatively small quantity. Such a factor means that even for minimal distances between bodies, our interaction forces do not explode to infinity, which would, unfortunately, cause instabilities in the simulation. Hence, the new gravitational interaction force will be of the form:

Fij = −Gmimj * (1/(rij^2+eps))*rij/|rij|,

where rij = ri-rj is the vector from body j to body i, eps is a small number (softening factor). The rest of the simulation is relatively straightforward and boils down to integrating the equations of motion, arranging bodies appropriately to account for the central and dominant mass, and visualization. In this type of simulation, primarily due to the principle of conservation of energy, Verlet's method is a good choice for numerical integration. So let us write down the algorithm for the planetary system simulation:

1 We are working on an 800 × 800 canvas.
2 Place a large mass M in the center of the array.
3 Arrange eight smaller masses with masses m1.m8<<M arranged farther and farther (equally spaced) from the central mass. Give them a small velocity tangent to a hypothetical orbit (e.g., perpendicular to the segment connecting the mass to the center of the system).
4 Integrate the equations of motion for all masses in the system using the Verlet method.
5 Assume that there is a law of universal gravity between bodies i and j, and the resulting force acting on both bodies has a different sign: Fij = −Gmimj * (1/(rij^2+eps))*rij/|rij|, where rij = ri-rj is the vector from body j to body i, eps is a small number (softening factor). Add the constant epsilon ("softening factor") to eliminate the problem of body collisions.
6 Visualize simply – black small circles on a white background.

Figures 47.1 and 47.2 are the results of the algorithm performed using a program generated from Prompt 47.

Task 1. Make a three-dimensional simulation of the planetary system with visualization.
Task 2. Design such an arrangement of three bodies (their masses, positions, and initial velocities) to realize a simulation in which the first orbits around the second, which again orbits around the third (along with the moon revolving around it). An inspiration here is to look at the masses of celestial bodies in the solar system.

DOI: 10.1201/9781003734062-48

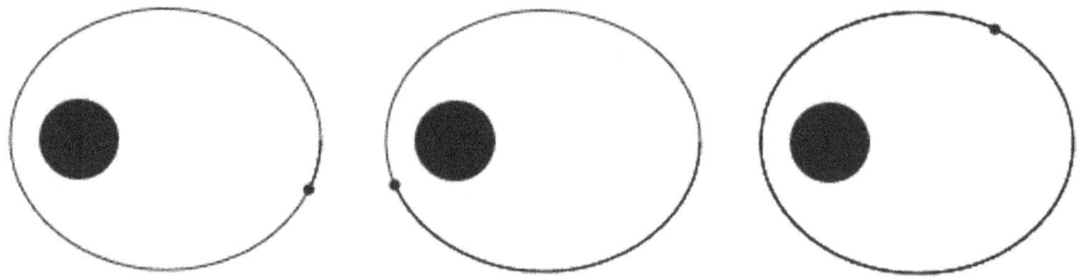

(a) (b) (c)

FIGURE 47.1 A simple system of two bodies of different masses (the larger mass here is 1000, the smaller 1), one of which orbits the other. The entry of the body with the smaller mass into the orbit of the one with the larger was made possible by setting an appropriate initial velocity (perpendicular to the vector connecting the two bodies of optimal size – not too large so as not to throw the body out of orbit, and not too small so as not to cause them to collide).

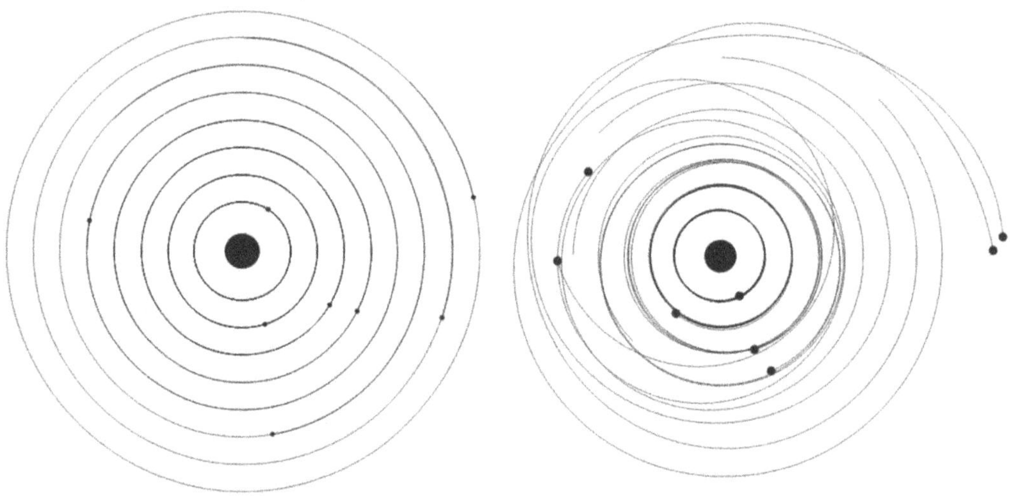

(a) (b)

FIGURE 47.2 Left: a planetary system of 8 masses stably orbiting the Sun positioned at the center of the system (the mass of the Sun is 50000 in our units, and the masses of the planets are m = 1). Right: similar simulation in the case of the Sun with a mass reduced by 10 times. Here, it is no longer able to keep its planets in a stable configuration, and even the Sun itself is moving due to its interaction with the planets.

48 Diffusive Tortuosity

Consider a random walk in two dimensions, focusing on the physical properties of the walk itself. If a particle performs a random walk, we can predict with a pretty good approximation how far (on average) it will move away from the starting point. This will depend, of course, on the mechanism causing this random movement and how strong and dynamic the forces are that move the particle. However, as a rule of thumb, the mean square displacement of the particle's distance from the starting point, when moving undisturbed by any obstacles, increases linearly with time with a coefficient known as the diffusion constant. This knowledge enables the prediction of how a substance composed of many such molecules, whose primary process is diffusion, will propagate. We will now test whether this is the case. On the grounds of simple modeling of random walks in two dimensions, we will examine whether the Smoluchowski-Einstein law is satisfied, linking the mean-square displacement of a particle that moves randomly with time by a simple linear relationship that can be represented as:

$$MSD(t)=2dDt,$$

where D is the diffusion coefficient, t is time, and the lowercase letter d denotes the spatial dimension. We will determine the magnitude of MSD(t) from a simple formula that averages the distances after random walks of independently moving wanderers:

$$MSD(t) = <|r(t) - r(0)|^2>=(1/N)*Sum(| ri(t)-r(0) |^2),$$

where sum runs over all hikers indexed by i, ri(t) is the position of the i-th hiker, and r(0) is the starting position of all hikers. The modeling itself is elementary, and in our edition, it will consist of the following steps: initializing a group of molecules at the starting position, performing random movement of each molecule in one of the available directions, tracking the trajectories of the molecules, and counting the averages to calculate the mean-square displacement.

It is interesting to study random motion and the diffusion coefficient in a system with obstacles. When a randomly moving molecule encounters an obstacle, its motion is slowed down, resulting in a decrease in the effective diffusion coefficient. To describe this phenomenon, a quantity called diffusive tortuosity is used, which is the ratio of the coefficient D0 in space to the coefficient D measured in a system with obstacles; i.e.,

$$T = D0/D.$$

Let us write down an algorithm that solves the random walk problem for both the empty system and the system with obstacles. The algorithm implements the problem in its entirety and also includes a procedure for fitting the Smoluchowski-Einstein theoretical formula:

1 Set N wanderers in the center of the system r(0)=(x0, y0).
  1.1 In the system, draw NB square-shaped obstacles of linear size A (over the entire available simulation area).
2 In a loop after time t:

DOI: 10.1201/9781003734062-49

3 In a loop after wanderers i:
   3.1 Take a step to one of the nearest neighbors ri(t) = ri(t – 1) + dr, where dr is a random offset in one of the four directions. If there is an obstacle at the target location, do not make the move.
   3.2 Draw the trajectory of the point ri(t).
   3.3 As you move, calculate on the fly the mean square distance of the simulation from the starting point MSD(t) = <|r(t) – r(0)|^2> = (1/N)*Sum(| ri(t)-r(0) |^2), where the sum runs over all wanderers indexed by i.
4 Draw the trajectories of all the wanderers.
5 Draw MSD(t) and numerical fit of the Smoluchowski-Einstein theory MSD(t) = 2dDt to the simulation data (fit by least squares method to MSD data obtained from simulation).

The following results of random walks and analysis of the mean-square displacement weighting were obtained using a program generated with Prompt 48 (Figures 48.1 and 48.2).

The above analysis shows that the so-called diffusive tortuosity in the studied system with obstacles is T=D0/D ≈ 1.69 (it is intuitively right). The number T should change with the degree of coverage of the system by obstacles, with extremes equal to 1 for an empty system and infinity for a large system close to the critical degree of packing by obstacles (percolation threshold).

Task 1. Investigate the average value of diffusive tortuosity <T> averaged over several systems with the same porosity but different configurations of obstacles. Draw a graph of the value of <T> from the porosity while marking where the percolation threshold of the system you are studying is. Try to find an empirical law describing the dependence of <T> on the porosity of the system (porosity as a ratio of pore area to the total area of the sample).

Task 2. Perform a random walk and diffusion coefficient analysis in a three-dimensional system with six allowed wanderer directions and proper visualization.

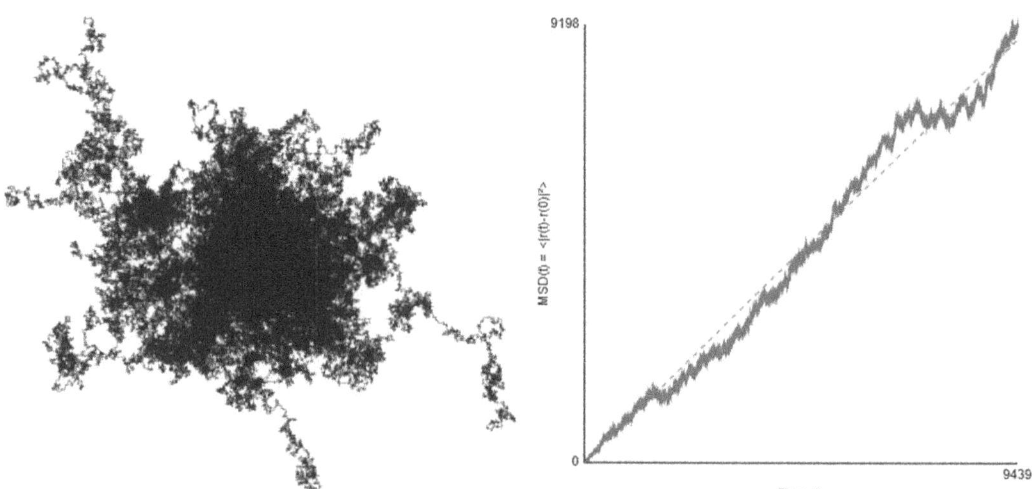

(a) (b)

FIGURE 48.1 (Left) Two-dimensional random walk of 50 independent wanderers starting from the center of the system (about 9400 steps). (Right) Fitting the mean-square displacement of MSD(t) to Einstein's theory, which in this case gave an approximation of the diffusion coefficient of D0≈0.236.

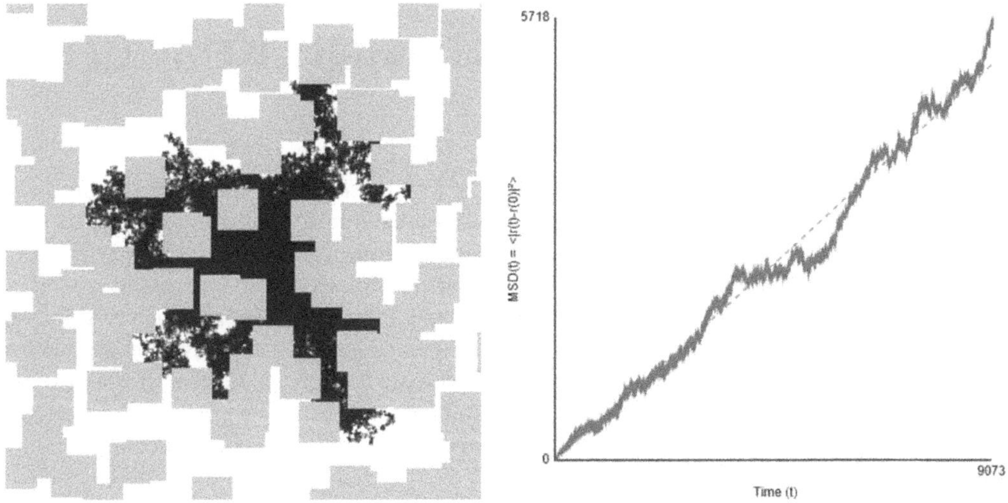

(a) (b)

FIGURE 48.2 (Left) Random walk of 50 wanderers in 2D in a layout with 200 obstacles arranged at random locations (about 5000 steps, visualization of part of the system). Right: a fit of Einstein's theory to MSD data from the simulation, resulting here in D ≈ 0.14.

49 Gray-Scott Diffusion Reaction

Diffusion with reaction is a fascinating phenomenon at the intersection of chemistry and physics, in which two (or more) different substances interact to cause the transformation of one into the other. Diffusion, as we have studied before, causes subtangents to penetrate each other, spreading out, and we have a pretty good understanding of what to expect here. However, incorporating reactions into the process allows us to simulate many interesting and unexpected patterns, such as stripes and blobs. Gray and Scott's model implements this phenomenon using two simple differential equations, which we will implement similarly to how cellular automata are solved. Readers interested in the origin of the original equations are invited to refer to Gray and Scott's original 1983 paper (Gray and Scott 1983) or browse the extensive literature that cites it. The reactions we are concerned with can be written as follows:

$$A + 2B \rightarrow 3B \; B \rightarrow P$$

They can be interpreted as follows: A and B react (first reaction). B transforms into an inert product P, which is removed from the system (second reaction). Additionally, we will consider feeding the system with substance A from outside (feed) and removing substance B. We will now write down the expressions for the local change in substance A and substance B, which we give after (Pearson 1993), which are consistent with the above reactions:

$$dA/dt = DA * \Delta A - AB\string^2 + F*(1 - A), \; dB/dt = DB * \Delta B + AB\string^2 - (F + k)*B,$$

where A and B are the concentrations of substances A and B, the derivatives on the left are taken after time, and the first words on the right with the constants DA and DB are the diffusion constants for substances A and B, respectively (the designation Δ is Laplacian). The constant F is the dimensionless rate of delivery to the system of the two substituents (in modeling terminology, it is referred to as the so-called feed rate). In the first equation, we take the expression F*(1-A) to avoid the problem of concentration exceeding the value of 1. In the second equation, the expression (F+k)*B is subtracted in the process in question and denotes the removal of substrate B both in the process of conversion to inert substance P and in the process of removal with a reaction rate proportional to the current value of B (with a constant F).

The expressions AB^2, which are respectively subtracted from the first and added to the second equation, are the model for the first reaction shown. It takes two molecules of substance B to convert A+2B→3B, so the reaction should occur less frequently for low concentrations of B. The model for this situation is the expression AB^2 (the value of B^2 for B<1 will be less than A). The reaction will cause a decrease in the number of molecules of substance A (hence the word AB^2 stands here with a minus) and an increase in the number of molecules of substance B (hence here the word AB^2 stands with a plus in the second equation). With this construction of the equations, in addition to the diffusion coefficients, two parameters (k and F) will appear in the model, allowing us to control the model's behavior and the resulting patterns. It now remains for us to transform the aforementioned equations into a form ready for calculation, and we will do so by introducing a differential scheme instead of the time derivative on the left-hand side and introducing a time step dt into the equations:

$$Anew = A + dt * (DA * \Delta A - AB\string^2 + F*(1 - A)),$$

$$Bnew = B + dt * (DB * \Delta B + AB\string^2 - (F + k)*B).$$

DOI: 10.1201/9781003734062-50

We left the Laplace operator Δ in this form, but keep in mind its expansion. We can, for example, use a five-point differential scheme, where eq.

$$\Delta A \approx A(x+1, y) + A(x-1, y) + A(x, y+1) + A(x, y-1) - 4A(x, y))/h\hat{}2,$$

where h is the size of the mesh, we are now ready to construct an algorithm that describes the entire model:

1 We work on W × H computational grids (initially 800 × 600).
2 We hold (scalar fields) A, Anew, B, and Bnew on the grids.
3 The initial values of A and B are: A = 1, B = 0 (in the whole area) with a slight disturbance in the central part of 20 × 20, where A = 0.5, B = 0.25.
4 After initialization, add a random value from −dr to dr.
5 In a loop, for each cell perform the operation:
Anew = A + dt * (DA * ΔA – AB^2 + F*(1 – A)), Bnew = B + dt * (DB * ΔB + AB^2 – (F + k)*B).
6 Use a 5-point formula to develop the Laplace operator Δ, e.g.

$$\Delta A \approx A(x+1, y) + A(x-1, y) + A(x, y+1) + A(x, y-1) - 4A(x, y))/h\hat{}2.$$

7 Visualize on the canvas. Show fields A and B using the field values to multiply separate palettes added together.

The result of running this algorithm from a simulation generated by the program using Prompt 49 is shown in Figure 49.1. This is a typical example of how the Gray-Scott model works with a gradient change in the values of k and F (here, the change in both values is performed by the program in the x-direction).

In the figure, we can see how the system generates characteristic stripes and how the phase transition to droplets occurs, followed by the disappearance of one of the substances (left-bottom part of the system).

Task 1. Run the same program with gradients separately for the x and y directions and generate a simulation in which F changes vertically and k changes horizontally by taking parameters from exemplary system from the Internet and compare it to your visualization.

Task 2. Generalize the algorithm into three dimensions and run a simulation in which one of the additional parameters (e.g., DA or DB) changes in the z-axis. Use the volumetric data visualization technique of your choice (e.g., using the marching cubes algorithm) to determine the iso-surface of concentration and draw it in 3D.

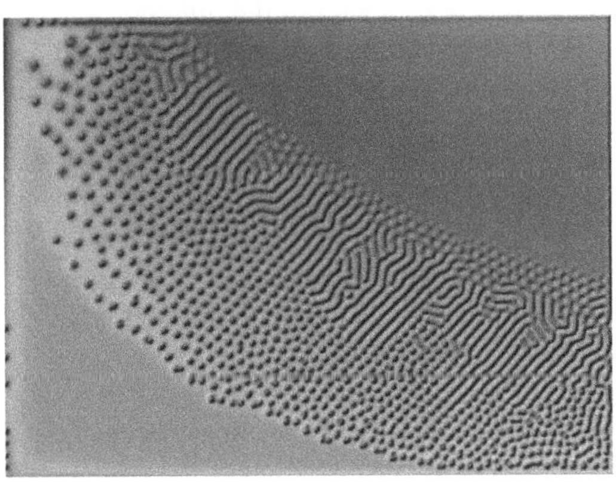

FIGURE 49.1 Result of a Gray-Scott simulation (diffusion with reaction with linear variation of the parameters k and F in horizontal and vertical direction).

50 Ant Colony

In this section, we will create a simulation of a large group of ants with a specific task: bringing food to the anthill. Our goal is to create a model of a complex system that organizes itself in order to perform the set task as efficiently as possible (see e.g. https://www.reddit.com/r/blender/comments/13oro3p/ive_added_a_bunch_of_new_features_to_my_ant/).

Mimicking real ants, we will consider the process by which ants leave a trail (Collett et al. 2025), which will be utilized for wayfinding throughout the community. After all, let us imagine one ant starting its movement from the anthill. Initially, it will move more or less randomly, turning in a random direction and walking ahead for more extended periods, essentially exhibiting a form of random walking; however, if this ant were to leave traces. Since it started its walk from the anthill, its footprints led right there. What if there are ants in the system that carry food? If they leave footprints, all you have to do is follow them to find the food!

We will assume that there are two types of ants, A and B, which, depending on their current state, either seek food (type A) or carry it back to the anthill (type B). This difference will result in different types of pheromones (FA and FB) that the ants leave behind. In the simulation itself, the A ants will pursue the FB pheromones and vice versa, and the pheromones themselves will diffuse over time (much like scent in the air) and fade over time. All this is done to ensure that the freshest traces are felt most by the ants.

Each ant will have some field of vision in front of it (at an angle and of a certain length), and it is in this angle that the appropriate pheromone, after-feed, or part of the anthill must be found to attract the ant.

Additional elements will be walls (caves) that will repel ants. We will also assume simple interactions between ants, based on elasticity, which will prevent ants from overlapping with each other.

The algorithm for the above model of the anthill and the collective movement of ants is as follows:

1 In the lower left corner of the canvas, W × H = 200 × 150, place a circle of radius W/8; this will be the anthill. In the anthill area, place N = 50 ants randomly.
2 The ants move in a fixed direction (each is assigned a velocity vector with a modulus of 1) and have a floating position (despite showing an ant on the grid, mathematically this object can have incomplete positions). Use periodic boundary conditions or conditions with reflection (parameter).
3 The ant makes a smooth, random rotation of the direction vector every so often. Let the user specify the rate (0, 1), range (45 degrees), and weight W (0, 1) of these changes.
4 The ant leaves a trail (hereafter referred to as a pheromone) behind it. Deposit it in the pheromone accumulation array (floating point number type) by adding the EPS value (parameter) where the ant is currently located. There will be two types of ants: going for food (type A) and returning to the anthill (type B). Hence, there will be two types of pheromones: FA and FB, which are left by the corresponding types of ants. Separately record the pheromone information of the respective ant type.
5 Pheromones disappear over time (exponentially or linearly, your choice) and are subject to diffusion (with a parameter – the diffusion constant).
6 In the upper right corner, on the grid, mark a circle with radius R = W/8. This is the food area.
7 The range of vision of the ant is limited to the angle hooked on its head, which collects information from before the ant (let the user change the dilation of the angle and the range of sensing).

DOI: 10.1201/9781003734062-51

8 If an ant in state A (going for food) has food in its range of vision, it smoothly turns toward it with a weight of W2 (1.0). When a field with food directly under its head is encountered, it intercepts it and then enters food-carrying mode, continuing to move while carrying a particle of food on its head. This particle also disappears from the food field. The ant, after capturing the food, begins a walk in which it smoothly turns 180 degrees, and then, already with a new direction, continues its further trip to the anthill.
9 Design the cave. To do this, set all cells that do not belong to a rectangle with dimensions (W-W/32, H-H/32) with rounded corners with SIGMA radius of curvature (dimensions and angle to be set by the user) to wall-type cells. No ant can enter the cave area (or any other obstacle).
10 If an ant senses a cave wall at an ODLSCIANA distance (let the user change this size) – rotate its direction of movement at a rate that will ensure a smooth rotation but also be fast enough to avoid the obstacle or move in a different direction (the parameter is the rotation rate).
11 If an ant carrying food finds itself in an anthill, it leaves food there, which disappears. The anthill counts and displays the amount of food brought to it in text. After leaving food in the anthill area, a new ant appears in a random vacant spot. After leaving the food, the ant starts walking in a new direction, turning 180 degrees, and then continues its trip for food.
12 The ant adjusts its angle of movement to follow one type of pheromone. Its direction vector changes toward the highest concentration of pheromones of the opposite type (ant A is attracted to FB pheromones and ant B is attracted to FA pheromones).
13 The ants avoid each other and do not overlap, but this should be implemented smoothly, without a rigid condition (use, for example, virtual springs that gently push the ants away from each other if they get too close to each other, where the user sets the spring constant).
14 If an ant in condition B (returning to the anthill) has an anthill in its range of vision, it smoothly turns toward it with weight W3.

At this point, instead of immediately stating the results of prompt 50, let us discuss one point. The above description of the anthill model is complete. However, several details that came out in the process of creating the prompt came out in so-called practice (e.g., the need for ants to turn back after taking food or leaving it, the need to determine the distance of interaction with walls, or details about the visualization of ants and pheromones themselves) came out only in the first attempts to create the model prompt. Indeed, the ant model proved to be a challenging task for language models. The degree of complexity, the individual behavior of the ants, the deposition of information (pheromones) on the grids, the handling of events (eating), the states of the ants, the details of animation and the physical model (repulsion, movement) – everything together is so complex that it took about 100 attempts, with almost every attempt modifying the prompt, to arrive at the solution and the prompt presented in this chapter. Moreover, even now, when the following algorithm is implemented, and we present the results in the pages of this book, it is an algorithm and process so complex that, fraught with the problem of temperature and deviation from the chosen path, language models can sometimes generate erroneous programs that do not cope with the model. However, we present both the algorithm and the prompt and results next, because we were finally able to generate a proper, working program and create a reasonably reproducible prompt.

How to deal with it? The process of creating the 50th prompt was one of the lengthiest, from the first prompt, written quickly, to a complete and working 50th prompt that took several days of hard work. One of its elements was the execution of trial prompts, which implemented parts of the algorithm detached from the rest. For example, the behavior of a single ant and its proper movement, including turning around on the walls of the cave, is a challenge in itself, and detailing it in the final prompt is very valuable. It can save nerves when trying to execute it. You certainly should not give up; the language model has a certain inertia and randomness, and sometimes, especially when you

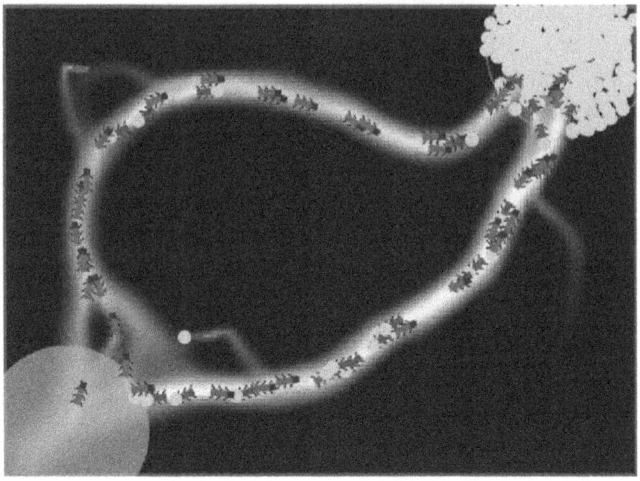

FIGURE 50.1 A colony of ants forms a connection between the anthill (lower left corner) and the food (upper right corner). The ants search for food and after collecting it – return to the anthill (with food attached). On the path formed between the anthill and the food, some ants walk back and forth, still leaving pheromones and strengthening the connection even more. The formation of the connection was spontaneous and self-organizing, and typically, each run of the program ends with a different configuration of paths between the anthill and the food.

are not too aware of what element is too poorly described and subject to essentially random prediction by the language model, you have to try to polish and run the Prompt until you get it right.

The result of the above anthill algorithm and the collective behavior of ants, executed with a program generated using Prompt 50, is given in Figure 50.1.

Task 1. Add removal of ants that do not find food for a specified period of time (parameter). Such ants should no longer exist in the simulation, nor should they leave pheromones.

Task 2. Perform a simulation in a maze. To do this, draw obstacles and corridors. Ants should solve the maze problem and find the shortest path connecting the entrance (the anthill) with the exit (the food). As an extension of your knowledge, read about optimization algorithms (https://en.wikipedia.org/wiki/Ant_colony_optimization_algorithms).

51 Wave Equation

Wave phenomena are ubiquitous. From waves on water, waves propagating on musical instruments, to sounds, which are also the result of propagating waves (in this case, density). The wave model we introduced earlier was very simplistic – it did not allow us to set wave properties, visualize, or add obstacles. Meanwhile, diffraction phenomena require the inclusion of obstacles, without which the systems that can be studied are limited. In this section, we will develop a model for simulating wave phenomena using the wave equation, a partial differential equation that describes the propagation of waves in time and space. This equation can be written as follows:

$$∂^2u/∂t^2=-v^2*(∂^2u/∂^2x^2 + ∂^2u/∂y^2),$$

where the partial derivatives of the pivot u(x, t) are taken in time (the second derivative on the left) and in space (the Laplacian on the right). This equation, in general, cannot be solved analytically. Here, we will use a technique I first encountered in D. Potter's book (Potter 1982; Matyka 2021), which involves introducing a new variable – the first derivative of the tilt, denoted by the letter E; that is,

$$E=∂u/∂t.$$

With the variable E, we can write the wave equation in simpler form, taking only the first derivative in time:

$$∂E/∂t = −v^2*∂^2U/∂t^2.$$

We can already solve the last two equations quite simply by using simple differential schemes, such as using formulas described on regular grids:

$$E(n+1, i) = E(n, i)+v^2 * lapl(U) dt$$

$$U(n+1, i) = U(n, i)+E(n+1, i)*dt,$$

where i is the grid node and n is the time step number, dt is the time step. In the 1D case, lapl(U) is the second derivative, which can be approximated by, for example, a three-point formula. In the 2D case, here we use the so-called differential analog of the Laplace operator, which can be, for example, a nine-point one (Matyka 2021). By performing an iteration of the above equations and applying appropriate boundary conditions to the values of U and its derivatives, we can conduct quite complex simulations of wave phenomena. The algorithm is not particularly complicated, although we will perform it taking into account an advanced element that we do not describe theoretically – we will assume that the language model will know how to find the appropriate procedure on its own. This is because it is about the attenuation of the wave at the edges of the system. The wave reflected from the edges disturbs the image of the simulated system quite strongly, leading to the formation of a standing wave quite quickly. A whole branch of modeling related to boundary conditions has arisen for this reason, the main task of which is to create, as it turns out, a non-trivial condition in which a wave incident on a rigid surface attenuates. It is like trying to create a material that absorbs a sound wave – a task not easy, even in the real world. There are such methods, such as PML (Perfect Matched Layer), in which a special layer of cells near the edges that gradually attenuate the incoming waves without generating

DOI: 10.1201/9781003734062-52

reflected waves (or generating them but to a much weaker degree), and the MUR method, in which an approximation of the solution in which the outgoing wave to an infinite array is applied in the simulation domain. For details, we refer you to the literature on the subject, for example (Xu et al. 2009). Here, we will indicate in the algorithm which methods we intend to use.

So let us write down the complete algorithm needed to simulate wave phenomena using the differential method introduced in this chapter:

1 Work on a W × H grid.
2 In the center of the array, place a point that will smoothly, according to the sin(t) function, change the amplitude.
3 The rest of the grid will be the computational grid.
4 After clicking start, numerically solve the wave equation in time:

$$\partial^2 u(r, t)/\partial t^2 = -v^2*(\partial^2 u/\partial^2 x^2 + d^2 u/\partial y^2),$$

where u(x, t) is the deflection field, the sign "^" denotes the derivative, and v denotes the speed of propagation of the wave.
5 Use the method of splitting the wave equation into two equations by introducing a new variable, i.e., (partial derivative)

$$E = \partial u/\partial t$$

which gives, after transforming the wave equation, its simpler form:

$$\partial E/\partial t = -v^2*\partial^2 U/\partial t^2$$

Discretize the above two equations for E on a numerical grid. Use a nine-point analog to discretize the Laplace operator from the wave equation, and discretize the time derivatives with a 1st-order differential analog. From E, you can count U (pivot) and draw the result, which amounts to following the procedure:

$$E(n+1, i) = E(n, i)+v^2 * \text{lapl}(U)\ dt$$

$$U(n+1, i) = U(n, i)+E(n+1, i)*dt,$$

with lapl(U) – the differential analog of the Laplace operator. Do not use suppression in the formulas.
6 On the edges, use boundary conditions with damping (choose several methods and give the user a choice, let MUR and PML be among these methods).

Figure 51.1 is the result of Prompt 51, which produces a program that simulates wave phenomena using the presented algorithm. The Prompt itself consists of the above algorithm supplemented by visualization issues and a user interface with the ability to draw obstacles and change the properties of the very medium in which the wave propagates. The solution is very general, and the result we show below is less than 10% of the capabilities of even this simple program, which was generated for this book.

Task 1. Generalize the problem to a three-dimensional one (directly modify the wave equation, where you need to add the derivative in space after the third component to the expression with a Laplacian). Then, solve the wave equation in 3D and create a visualization using an algorithm to display such a field as a cloud of transparent points (you may need to

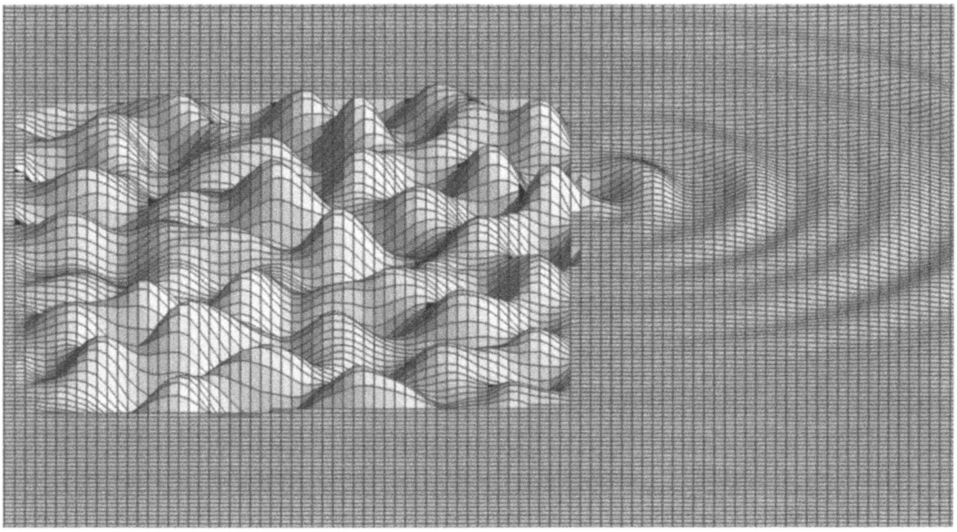

FIGURE 51.1 Standing wave (left) and diffraction of the wave on the slit (right). A small slit has been added in the top part (roughly ¼ from top) of the right vertical wall bounding the container in which the wave source produced a standing wave. The simulation was based on a program that utilizes an algorithm for the discretization of the wave equation. The applied boundary watts (MUR) worked very well here, damping the wave behind the slit after it reached the edges of the system.

use a rendering technique similar to ray tracing and cloud-type data). Simulate two wave pulses in 3D and the phenomenon of interference.

Task 2. Add the possibility of moving the pulse by moving it during the simulation and conduct a simulation in which the pulse moves faster than the wave. Observe how the pulse generates a shock wave as it moves in a straight line.

52 Electron in a Magnetic Field

A particle with an electric charge q moving with velocity v in a magnetic field B is affected by the Lorentz force:

$$F = q^*(v \times B),$$

where the sign of x denotes the vector product. Knowing this force allows you to use the classical equation of motion:

$$d^2 r/dt^2 = F/m.$$

The equation of motion for a particle can be solved using one of the known numerical methods. The difference, for example, from the simulation of planetary motion introduced earlier, is that the Lorentz force is used instead of gravitational forces. At this point, the Verlet method, which preserves energy in the system over extended simulation times, is the most suitable choice. At this point, it is also worth noting that we will be working in dimensionless variables, where the charge of the particle (here, the negatively charged electron) will be −1. However, carrying out simulations in a system with a directly given magnitude of charge q is very difficult, because the standard for writing floating-point numbers does not well support numbers on the order of 10^−19, and such calculations would generate significant numerical errors. Therefore, it is essential to note that we are performing calculations with an elementary charge of q = −1, which will enable us to generate and analyze the trajectories of electrons in magnetic fields of varying structure, our primary objective. So let us write down the basic minimum algorithm that implements the simulation of the electron model in a magnetic field:

1. We work in an 800 × 600 layout.
2. An electron (q = −1) is started from the point x0, y0 with an initial velocity V0 = (V0X, V0Y, V0Z).
3. There is a constant magnetic field B = (BX, BY, BZ) in the system; initially B = (0, 0, 1).
4. Let us write down the equation of motion of the electron: d^2 r/dt^2 = F/m, where F is the force acting on the electron, m is its mass, and the second derivative on the left side stands for acceleration.
5. Integrate the equation of motion of the electron, taking into account the Lorentz force (interaction of the moving electron with the magnetic field), i.e., F = q*(v x B), where q is the electron's charge, v its velocity, and B the magnetic field. Use the Velocity Verlet, constant-energy method.
6. Visualize the motion of the electron and the path of motion.

In the following section, we present one of the results generated using the above algorithm and Prompt 52, which was built on it. This is a simulation of a system of 10 electrons, initially arranged at the same point with a slight random scattering of their initial velocity, whose motion is carried out in the presence of a magnetic field with a complex gradient structure. It should be noted here that the electrons are independent here; that is, they do not interact with each other, but the purpose of the simulation was not to count the collective motion of the electron system, but to visualize how the trajectories of independent, charged points that take off from the same point at different speeds differ. It is also worth noting that Prompt 52 includes a minor refinement of Verlet's method relative to the algorithm above, and comments to examine the numerical errors of the procedure, as the

DOI: 10.1201/9781003734062-53

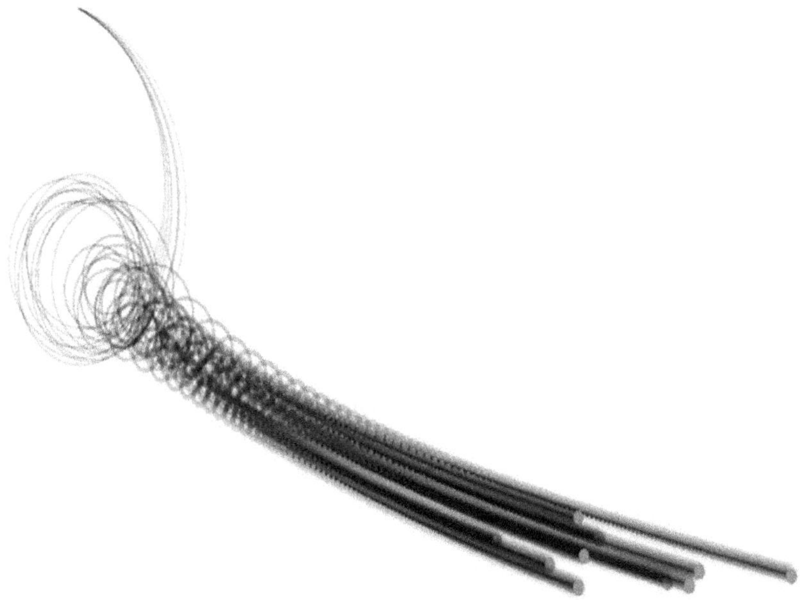

FIGURE 52.1 A group of 10 independent electrons moving in a magnetic field that is variable in space (linear gradient separately for each component). The electrons start from the same point, but at minimally different speeds. The thickness of the line reflects the motion of the electron in the z-axis. Parameters: dt = 0.01, N = 10, x0 = 70, y0 = 50, z0 = 0, DR = 0, V0X = 60, V0Y = 20, V0Z = 0, DV = 4, BXmin = 0, BXmax = 1 (x gradient), BYmin = 0, BYmax = 0.25 (y gradient), BZmin = 0, BZmax = 1 (z gradient), track length = 21,200.

language model initially lacked accuracy, causing the electron trajectories to break down due to numerical errors (Figure 52.1).

Task 1. Make a 3D visualization with the ability to rotate and view the generated electron trajectories from anywhere in the system.

Task 2. Add the electric field E to the system and perform simulations in a system where it is constantly directed in one direction. To do this, write the equation for the Lorentz force in general form with E and B fields. How does the inclusion of the E field modify the trajectories of the electrons?

53 Smoothed Particle Hydrodynamics (SPH)

Is it possible to effectively simulate the behavior of a fluid by tracking the motion of its elements interacting with each other? Is it possible to simulate water in a glass in real time, such as for a computer game? There is an approach that allows it. No complicated formulas for derivatives, no special algorithms for solving differential equations, just particles, forces, and moving them in space using the simplest Euler method.

The way of describing a hydrodynamic system can be divided into two main classes. One description is based on the spatial description of the fluid properties, such as on grids on which the physical fields describing the fluid are described (the so-called Euler approach). The other is a description of moving fluid elements (e.g., molecules or groups of molecules), known as the Lagrange approach. In this section, we will examine the second approach and describe the Smoothed Particle Hydrodynamics (SPH) model. This simulation method is notable for its interpolation-based approach, which avoids the use of a computational grid. We will base our solution on the work of Auer (2009) and Ihmsen et al. (2014).

The basis of the method under discussion is interpolation. Suppose there are N points in space with some property A (it can be density, color, but also velocity component, etc.). We know the values of A(xi, yi), where xi and yi are the positions of all points in the system. To calculate this property at any other point (x, y) in the space, we use the interpolation formula:

$$A(r) = \Sigma j\ Aj^*(mj/\rho j)W(r\text{-}rj,\ h),$$

where the sum runs over the particles in the vicinity of the point r of radius h, rj is the position and mj is the mass of the j-th particle, ρj is its density, and W(r-rj, h) is the so-called interpolation kernel – a special function, normalized to one, which generally tends to 0 for large distances r-rj and has a maximum at 0. An example formula for the W function is a polynomial formula, here given with one parameter:

$$W\_poly6(d,\ h) = 315/(64^*\pi^*h\hat{\ }9)^*(h\hat{\ }2 - d\hat{\ }2)\hat{\ }3 \text{ for } d< = h \text{ and}$$

$$W\_poly6(d,\ h) = 0 \text{ for } d>h,$$

where d = r-rj is the distance between two SPH particles. The choice of the W function is not straightforward, and it is common in the literature to find different functions depending on the context and the quantities it is expected to approximate; see for example (Auer 2009; Ihmsen et al. 2014). We will use two functions, in addition to the polynomial function above. These two functions differ in shape and differentiation properties (see algorithm) and have been selected from the literature based on observations of their desirable characteristics for different forces and magnitudes.

The above formulation of interpolation using W kernels also provides the possibility of determining analytical formulas for the derivatives of quantities at a given point, such as the gradient. This is because instead of performing the spatial derivatives of the calculated values, we perform them directly on the interpolation formula:

$$\nabla A(r) = \Sigma j\ Aj^*(mj/\rho j)\nabla W(r\text{-}rj,\ h).$$

 DOI: 10.1201/9781003734062-54

In the above formula, W can be differentiated manually and analytically (which we will also do to simplify calculations). We can similarly introduce other differential operators.

For the SPH method, we will use the Navier-Stokes equation in molecular form (Auer 2009):

$$\rho(ri)\ dvi/dt = -\nabla p(ri) + \eta\nabla\nabla v(ri) + \rho(ri)g,$$

where the derivative on the left refers to the acceleration of a single fluid particle, and the successive expressions on the right correspond to pressure, viscosity, and gravitational acceleration. The Navier-Stokes equation in this edition no longer contains a continuity equation, because the number of particles and their mass are constant. As an aside, we are discussing a particle, not a molecule, because we are not working at the molecular level in SPH but rather treating a piece of the fluid as a larger collection of particles.

Our SPH fluid simulation will therefore consist of particles distributed in space, which will interact with each other through the forces described by the expressions in the Navier-Stokes equation above. To solve the above equation for each particle in the system, we will first calculate the density for each particle in the system by interpolating the mass across its neighbors:

$$\rho(r) = \Sigma j\ mj\ W(r\text{-}rj, h).$$

Next, we determine the forces occurring on the right side of the Navier-Stokes equation, using formulas (see Auer 2009 for a deeper explanation of the origin of the formulas) for pressure:

$$-\nabla p = Fp = -\Sigma j\ (mj{*}(p{+}pj)/(2{*}\rho j)) * \nabla W\_spiky(r\text{-}rj, h),$$

where we calculate the values of p and pj (denoted interchangeably as pn in the following algorithm) from the equation of state:

$p = k{*}(\rho - \rho 0)$, where $\rho 0$ is the density of the fluid at equilibrium.

Similarly, we write down the viscosity term:

$$\eta\nabla\nabla v(ri) = \eta\Sigma j\ mj\ {*}(vj\text{-}vi)/\rho j\ \nabla\nabla\ W\_viscosity(ri\text{-}rj).$$

After calculating the forces on the right side of the Navier-Stokes equation, the equations of motion can be regressed (even the simplest Euler integration method or the higher-order LeapFrog method can be used for this). We will write down this simpler method of integration:

$$v = v + a * dt$$

$$x = x + v * dt,$$

where the acceleration a is calculated directly from the Navier-Stokes equations using the above procedure.

Readers interested in learning more about the SPH method, its advantages and disadvantages, and tracing its development are invited to read one of the key papers on the subject (Monaghan 2005). We are now ready to write down the general algorithm, which will be supplemented by the XSPH procedure for velocity smoothing, in which, after counting the forces, a correction is made for each i-th particle before updating the position:

$$vi = vi + eps{*}\Sigma j\ (mj/\rho ij){*}(vj\text{-}vi){*}W(r, h),$$

$$\text{where } \rho ij = 0.5{*}(\rho i{+}\rho j).$$

Let us write down the basic SPH algorithm:

1 We work in a 2 × 1.5 (W × H) arrangement, where (0, 0) lies in the lower left corner.
2 Distribute NX uniformly over NY points in a rectangle (0, 0) – (W/4, H/2) with additional random scatter dx, dy.
3 Define the interpolation kernels and their gradients (the "^" sign is a power-up):

W_poly6(d) = 315/(64*π*h^9)*(h^2 – d^2)^3 for d<=h and

W_poly6(d) = 0 for d>h,

W_poly6_grad(d) = 945/(32*π*h^9)*(h^2 – d*d)^2,

W_poly6_grad_x(rx, ry, d) = –rx * W_poly6_grad(d),

W_poly6_grad_y(rx, ry, d) = –ry * W_poly6_grad(d),

W_poly6_lapl(d) = 945/(8*π*h^9) * (h^2 – d^2) * (d^2 –3/4*(h^2–d^2)),

W_spiky_grad(d) = 45/(π*h^6*d) *(h-d)*(h-d),

W_spiky_grad_x(rx, ry, d) = –rx * W_spiky_grad(d),

W_spiky_grad_y(rx, ry, d) = –ry * W_spiky_grad(d),

W_viscosity_lapl(d)= 45/(π*h^5) * (1 – d/h).

In kernels, ensure secure values of d = 0 or d close to 0 to prevent kernel values from exploding.

4 For each particle i, calculate its density by calculating the sum over its j neighbors for which r<=h. To do this, calculate the expression ρi = sum(j) m*W_poly6(r, h), where r = |ri – rj| (also consider the contribution of the ith molecule itself).
5 For each i-th molecule, clean the force acting on the molecule and the gradient and Laplacian accumulator, and then make a loop over its j neighbors (for which r<=h) to add the forces from that molecule and calculate the surface tension properties:
   5.1 Constrain ρ and ρn to a maximum value of ρ0, i.e., ρ = max(ρ, ρ0) and ρn = max(ρn, ρ0). Calculate the pressures: p = k*(ρ – ρ0), pn = k *(ρn – ρ0), where ρ and ρn are the density of the particle and its neighbor.
   5.2 Add the pressure force Fp = –(m*(p+pn)/(2*ρn)) * W_spiky_grad(r) (use separate versions for the x and y components of the gradient).
   5.3 Add the viscous force Fvisc = (η*m*(vn – v)/(ρn)) * W_viscosity_lapl(r), where v and vn are the velocities of the particle and its neighbor, respectively.
   5.4 Add gradient: (m/ρn) * W_poly6_grad(r) to the gradient accumulator (use separate versions for the x and y components of the gradient).
   5.5 Add laplasian: (m/ρn) * W_poly6_lapl(r) to the laplasian accumulator.
6 Total force F = Fp + Fvisc.
   6.1 Acceleration a = (F/ρ) + g
   6.2 Velocity v = v + a * dt
   6.3 Position x = x + v * dt
7 Default values of the variables h = 0.02, ρ0 = 10, η = 10.0, gy = –1.0, dt = 0.0025, k = 8.0, particle mass m = 0.003, random dx = 0.00001, random dy = 0.00001.
8 At walls and obstacles, reflect particles at a distance of DS=4*h (soft particle radius) and consider this distance when generating a column of points so that they are not generated in the collision area.

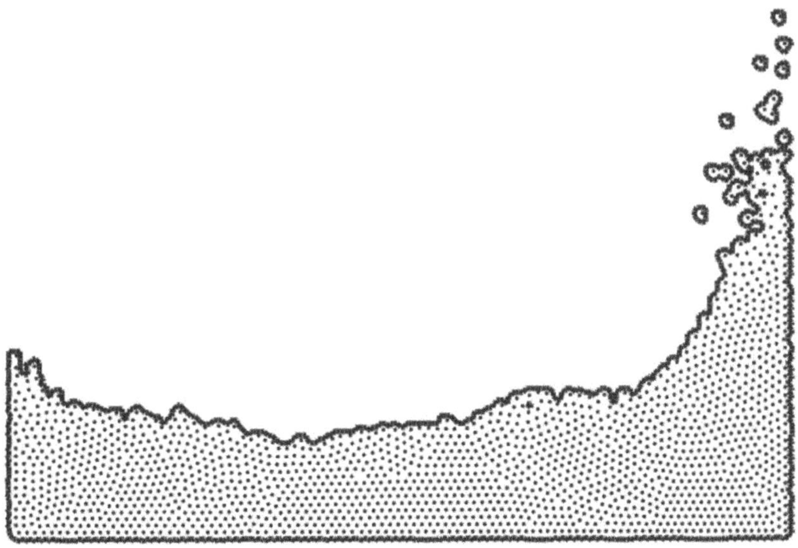

FIGURE 53.1 The behavior of a fluid with a free surface simulated by the SPH method. Initially, the fluid column was located on the left side of the system. Hence, the larger splash and resulting droplets on the right side, where the column of water hit with momentum. The visualization here shows both the particles and the contours of the iso-surfaces from the marching cubes (marching cubes) algorithm, which allows you to visualize the fluid surface formed from the particles.

9 Implement an additional velocity smoothing model in which, after counting the forces, a correction is made for each i-th particle before updating the position:

vi = vi + eps*sum(j) (mj/rhoij)*(vj-vi)*W_poly6(r, h),

where rhoij = 0.5*(ρi+ρj), and the sum after j runs on the neighbors of molecule i (for which r<=h).

Figure 53.1 is the result of a program generated using Prompt 53, which supplements the above algorithm with a description of the user interface and visualization, including the generation of fluid surfaces (using the marching cubes algorithm).

Task 1. Add the surface tension forces described using the additive so-called "color fields" in (Auer 2009) and perform simulations with its various parameters.

Task 2. Implement the LeapFrog method and investigate the stability and speed of the SPH method for larger time steps (dt).

Task 3. Generalize the solution to 3D and implement the model with shading programs (preferably use computational shaders for this purpose). Visualize the falling column in 3D, taking into account the rendering of the surface generated by the marching cubes algorithm, shadows, light refraction on the surface, and lighting.

54 Hydraulic Tortuosity (Research)

Is it possible to conduct scientific research using prompts to language models? Is it possible to ask questions, test, and verify hypotheses? In this chapter, we will construct an algorithm to generate a program that simulates flow through a porous medium, aiming to calculate its macroscopic parameters, such as tortuosity and permeability. These two quantities are the main characteristics of porous media that are studied in the context of fluid transport. The tortuosity T indicates how elongated the trajectories of particles are as they flow through a porous medium. Let us take one such particle as an example. If it were flowing in a pipe, without any obstacles, its path Le would be a straight line (Le = L) through the entire system. If there are obstacles, for example, in the porous medium filling such a pipe, particles bypassing the obstacles (or flowing through channels) will have to take a longer path (then Le>L and hence T>1). We can define the tortuosity of a single line as:

$$T = Le/L.$$

In the case of a large number of particles, one can, of course, take the average tortuosity from many such lines, which, however, is not numerically straightforward. For low velocities, which we will be concerned with here, it is much simpler to use a formula in which it is enough to make an integral of the velocity field over the entire pore area. In this way, the tortuosity value can also be determined:

$$T = <v>/<vx>,$$

where <v> is the average of the velocity modulus over the entire velocity field in the pore space, and <vx> is the average of the x-component. As shown in (Duda et al. 2011), among others, such a procedure leads to determining the macroscopic tortuosity of transport in a given medium, consistent with calculations based on streamlines. The magnitude of T is worth knowing for many reasons, as it not only allows for the unambiguous characterization of the medium but also facilitates the easier estimation of permeability (see Kozeny-Carman law).

In order to determine the permeability, the second parameter often studied in these types of problems, Darcy's law can be used directly:

$$q = -k/\mu * (\nabla P + \rho * f),$$

linking the pressure gradient and external force $\nabla P + \rho * f$ to the flow rate q through permeability k and viscosity μ. Straightforwardly, k can be determined from this law, since both q and the rest of the variables on the right-hand side are known from the solution and the data entered into the fluid solver.

In our example, we will generate systems with random obstacles that can overlap, resulting in different porosity depending on their number. The more obstacles we add to the system, the lower the porosity will be.

We will perform a simulation of fluid transport by solving the Navier-Stokes equations in the pore space of our system (we will use the lattice gas Boltzmann method introduced earlier for this), and on the velocity field results, we will perform the necessary tortuosity and permeability calculations.

DOI: 10.1201/9781003734062-55

The novelty of the algorithm we will develop will be the whole mechanism of testing multiple systems, averaging the results and constructing whole sets of results, hence the request to perform a whole series of measurements with simultaneous generation of systems of different porosity, averaging and calculation of standard errors from the measurement series, which we will have to implement to be able to track the statistical properties of our systems.

Our problem will be interesting as we will explore the possibility of studying systems that have not only overlapping obstacles, as in the work of Matyka et al. (2008) but also systems with a certain minimum distance between obstacles. These systems have different properties regarding the percolation threshold, but both will be somewhat similar in high porosities. Our goal will be to compare permeability and tortuosity results for both types of porous systems.

At this point, it is worth asking some open-ended questions – questions not answered in the literature, or questions or statements we can answer or verify with a potential program generated by our algorithm. These may include, for example, the following hypotheses:

1) The nonlinear nature of the tortuosity function from porosity in the 2D case presented in (Matyka et al. 2008) is a clustering effect of overlapping obstacles and will become linear for isolated obstacles. For non-overlapping obstacles, each new obstacle introduces a linear contribution to the tortuosity.
2) For isolated obstacles, we expect that from the first obstacle added to the system, its contribution will be linear and the function describing the twist will be of the form $T(\varphi) = a*x+1$ and will pass through the point $T = 1$ for $\varphi = 1$.
3) The tortuosity for overlapping obstacles is generally greater than that for systems composed of isolated obstacles.
4) The tortuosity for isolated obstacles does not explode to infinity (no percolation threshold).
5) The relationship $T(\varphi)$ is only quantitatively different for different shapes of obstacles.

Now, let us write a minimal algorithm that implements the simulation and measurement of tortuosity and permeability in porous media, in the context of performing calculations under the research hypotheses posed earlier:

1 Write an implementation of the Lattice Boltzmann method that simulates fluid flow in a channel with periodic boundary conditions on the x and y axes.
2 Accelerate the fluid by gravity acting in the horizontal, x-direction.
3 Place random obstacles that have a no-slip condition (velocity = 0) on the edges. Add a second deposition algorithm, in which the obstacles do not overlap and have a minimum edge distance of A.
4 Visualize the flow with 4000 massless particles following the velocity field.
5 When calculating, calculate the tortuosity and permeability.
6 Allow the user to vary the external force.
7 At each time step, recalculate whether the steady state has been reached.
8 Add a button to randomly place obstacles and a slider to change the overall porosity of the configurations generated with this button. Add the ability to change the parameter A – the minimum distance between two obstacles.
9 Allow control of the delta time used for time integration of massless particles.
10 Set the size of the calculation grid to NXxNY (300x200). Allow changing NX and NY.
11 Add a canvas next to the simulation with a graph of the tortuosity measured over time. This graph should reset every time the user presses the random configuration button.
12 Make sure the simulation is stable.
13 Add a text box with 20 lines.
14 Add a "save" button that adds data to the text box: simulation step, porosity, tortuosity, permeability, convergence criterion, and Reynolds number.

15 Add the possibility of a series of measurements in which we set the range and step of variation of porosity, the number of simulation steps to be performed, and the number of repetitions for one porosity NPOR.

In Figures 54.1 and 54.2, we show the results of the simulation run by the program generated by Prompt 54. The prompt was extended with some details on LBM method and deeper details on the computation procedure. In Figure 54.1, we have velocity fields represented as a color map, with obstacles and the instantaneous configurations of particles plotted in the fluid. We show two results

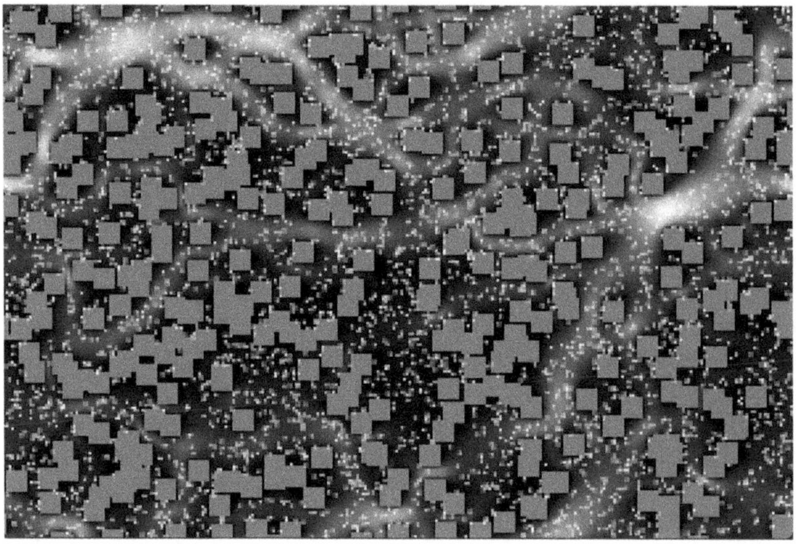

(a)

(b)

FIGURE 54.1 Example of velocity field in two systems with porosity $\phi \approx 0.7$ a) overlapping and b) separated (A = 4 l.u.). Both systems calculated for fx = 0.005, 320 × 200 layout, and 4 × 4 l.u. square obstacles. The calculated tortuosity is T = 1.31 and T = 1.15, respectively.

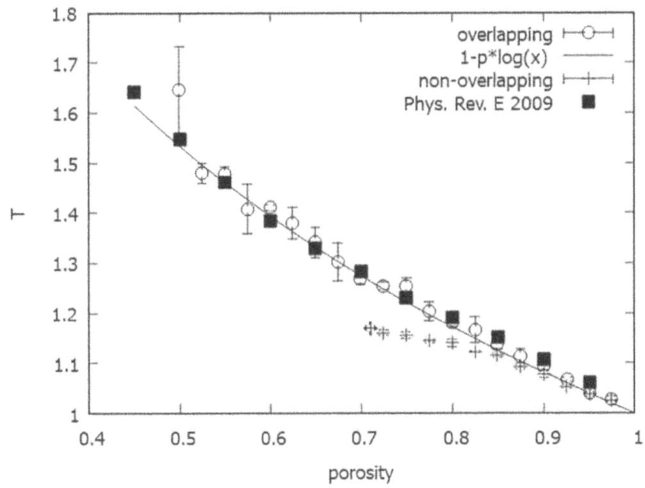

FIGURE 54.2 Tortuosity measurement in a system of overlapping and non-overlapping square obstacles. Measurement made for size 300 × 200, external force fx = 0.0005. In addition, the graph shows results from the work of Matyka et al. (2008) made for overlapping squares (filled symbols) and empirical formula derived from fits of the function to the data (solid line).

here to illustrate the difference between systems with the same porosity but different ways of distributing obstacles (with or without overlapping).

Next, we present an analysis performed with the same program resulting from Prompt 54, utilizing the functionality to perform a series of simulations. Thanks to this design of the program, we were able, in principle, with a few clicks, to perform tests of the T(φ) function for overlapping quadratic obstacles and repeat some of the results from (Matyka 2009). We then run simulations for non-overlapping obstacles (same graph next), which clearly show the differences in the hydrodynamic properties of the two systems.

Based on the previous results, with respect to the hypotheses posed earlier, we conclude that:

1) It is impossible to explain the origin of the nonlinear character of T(φ). Intrinsic anisotropy of random-type media probably dominates the shape of T(φ) due to periodicity (see Koza et al. 2009). The system of non-overlapping obstacles has a limitation from below on porosity (no clogging threshold – there is always a flow path here).
2) Our results do show linear contribution of single obstacles to tortuosity. The T(φ) approaches linearity close to $\varphi = 1$.
3) The tortuosity for overlapping obstacles is greater than that for systems composed of non-overlapping obstacles at lower porosities. The minor statistical errors for non-overlapping obstacles are striking.
4) The tortuosity for non-overlapping obstacles does not go to infinity (no percolation threshold).
5) In the range $0.9<\varphi<1$, both cases have similar character, which is mainly due to similarity in geometry (there is not much difference between overlapping and non-overlapping systems for only a few obstacles).

Task 1. Perform the same tests and measurements in the 3D version with particle visualization (without visualization of velocity fields). Draw the obstacles as transparent small cubes. Calculate tortuosity dependence on porosity and compare 2D to 3D results for both cases of overlapping and isolated grains. Test the hypothesis that in the 3D case, the contributions to tortuosity from each new obstacle are independent of their degree of coverage and are always linear.

Task 2. Develop a theory to graph the tortuosity dependence of porosity for the case of non-overlapping obstacles.

Task 3. Use other measures of tortuosity (the elongation of the current line measured from the tracks of the particles, the integral from the local tortuosity of the streamline, etc.) and study their behavior in relation to porosity.

Task 4. How will the tortuosity change if we attach the phenomenon of diffusion to the flow? Draw some graphs of T from porosity for different diffusion coefficients (different Péclet-a numbers).

Task 5. On the same obstacle systems, study the diffusion tortuosity and hydrodynamic tortuosity. Please point out the differences and examine whether they are different depending on whether you are working with an overlapping or non-overlapping obstacle system.

Task 6. Add to the project the ability to calculate and draw the velocity distribution function (the probability of finding components or the modulus of a specific value of velocity in the pore space). Try to identify the main differences between systems of different natures (overlapping obstacles or isolated). Answer the question: What is the reason for the reduction in tortuosity for isolated obstacles?

Task 7. Implement a stable LBM method (MRT, TRT, or any other relaxation or stabilization LBM method for high velocities) to determine the dependence of the tortuosity on the Reynolds number over a wide range. Investigate how the fluid structure changes as a function of the applied force. Try to reach the limit of your model for a single sample of your choice and perform several trials for different schemes to choose the one that will allow you to study the tortuosity from porosity for several extreme Reynolds numbers.

Task 8. Investigate the effect of y-axis boundary conditions (periodicity vs. no-slip conditions) on the tortuosity in the system for porosity $\varphi<0.8$. Do the results differ significantly? How could these differences be explained? What physical feature causes different boundary conditions to lead to different results? For a hint; see (Koza et al. 2009). Do the results affect our answer for hypotheses posed in the project?

Task 9. Instead of overlapping obstacles, build a channel system, a branched network of small randomly connected channels of random width WA with some random (user-adjustable) variation. Please ensure there is no slippage (velocity = 0) at their boundaries. The channels can be adjusted by the user, who can adjust the width distribution, sizes, connections, and all other possible properties of the network, including the porosity of the system. Study the tortuosity from the porosity of such a system and see if it has anything to do with the transport of fluids through media made of obstacles.

Task 10. Please, check how the location of kinetic energy (PI number) changes depending on the type of packing of the figures. Intuition tells you that it should differ between systems with a different type of packing, but the same porosity.

55 Shallow Water

Performing a complete simulation of fluid behavior is complex and computationally expensive. Suppose we do not need information about the fluid's behavior in its entire volume, and we are only interested in the behavior at the surface. In that case, there are models of so-called shallow water, which significantly simplify the issue by reformulating the flow equations (Chen and da Vitoria Lobo 1995). These models offered, many years ago, the ability to achieve real-time velocity in water simulation, which is necessary, for example, for computer games (Kass and Miller 1990; Lander 2000).

In this section, we will look at the model described in Kass and Miller (1990). In this model, we solve a one-dimensional wave equation where, in addition to the fluid's surface on which the wave propagates, a ground level exists whose height modifies the speed of wave propagation. This is an exciting idea, which, in combination with an algorithm that allows us to add sources and so-called fluid sinks (places where the fluid disappears), will allow us to simulate a range of water flow phenomena – from waves on the surface, waves hitting the beach, rivers, and even droplets moving on the surface.

The equation derived by Kass and Miller (1990) is the wave equation with modified velocity:

∂^2 h/∂ t^2 = gd * ∂^2 h/∂ x^2,

where h(x, t) is the height of the water column, x is the position in the horizontal axis (the problem is one-dimensional), t is the time, g is the earth's acceleration, d is the depth of the reservoir calculated from the difference between the height of the water column and the height of the earth's surface b(x, 0) over which this column is located, i.e. d(x) = h(x) – b(x). This equation describes the propagation of a wave with velocity sqrt(gd).

The solution proposed in the work under discussion involves using the discrete form of the equations, where the discretized functions d(x, t), h(x, t), and b(x) are stored on a one-dimensional computational grid. To solve the above wave equation, we start from some initial moment at which we know the above functions for the first two time steps (there is some freedom of choice here, and the final solution depends on it). We assume that b(x) is not time-dependent, and in the algorithm, we will indicate this by writing b(x, 0) – the time-independent value set as the initial state.

We discretize the above wave equation using classical finite differences on a numerical grid in one dimension. The complete derivation in (Kass and Miller 1990) includes both the introduction of the differential schemes of the fundamental derivatives in the above equation and the linearization process of the equation that leads to the final matrix formula. Following Kass and Miller (1990), we will use the viscosity version:

A*h(i, t) = h(i, t–1) + (1–τ) * (h(i, t–1) – h(i, t–2)),

where τ ranges from 0 to 1 and represents the viscosity of the fluid. The matrix A is a three-diagonal matrix, where on the diagonal lie the elements e0, e1, e2, . . ., en–1. In the diagonals lying above and below the central diagonal lie the elements f0, f1, . . ., fn–2. The elements are calculated from the formulas:

e0 = 1 + g * (dt*dt) * (d(0, t) + d(1, t))/(2*dx*dx), and

ei = 1 + g * (dt*dt) * (d(i–1, t) + 2*d(i, t) + d(i+1, t))/(2*dx*dx), for (0 < i < n–1), and

en–1 = 1 + g * (dt*dt) * (d(n–2, t) + d(n–1, t))/(2*dx*dx), and

fi = –g*(dt*dt)*(d(i, t) + d(i+1, t))/(2*dx*dx),

DOI: 10.1201/9781003734062-56

that come from the discretization and linearization of the wave equation introduced at the beginning of this section.

Iterating the above matrix equation will generate solutions for the behavior of the fluid in the shallow tank approximation. The complete algorithm is as follows:

1 Work in one dimension in a system of width (number of nodes) n.
2 Let u(x, t) be the column velocity of the fluid at point i at time t.
3 Let h(x, t) be the height of the fluid column.
4 Let b(x, 0) be the height of the substrate (Earth's surface) (constant over time).
5 Determine the initial moment h(x, 0), h(x, 1) and the shape of the substrate b(x, 0).
6 Let h(x, 0) be a straight line of value H/2 with an impulse of amplitude equal to the height of the system/10.
7 Let h(x, 1) be a straight line of value H/2 with an impulse of amplitude equal to the height of the system/10.1.
8 Let b(x, 0) be a straight line with a value of H/12.
9 Start the calculation loop after times (start with j = 2 with incrementation by 1):
10 If there are any sources or sinks of water in position i, add or subtract water from the current value of h(i, j).
11 Calculate for all columns and values d(i, j) = h(i, j−1) – b(i, 0), if d(i, j)<0, then d(i, j) = 0.
12 Calculate the new value of h(x, j) from h(x, j−1) and h(x, j−2) from Eq:
A*h(i, t) = h(i, t−1)+(1−tau)*(h(i, t−1)-h(i, t−2)), where tau is a value from 0 to 1 and represents the viscosity of the fluid. Matrix **A** is a three-diagonal matrix, where on the diagonal lie the elements e0, e1, e2, . . ., en−1. In the diagonals lying above and below the central diagonal lie the elements f0, f1, . . ., fn−2. The elements are calculated from the formulas:

e0 = 1 + g * (dt*dt) * (d(0, t) + d(1, t))/(2*dx*dx), and

ei = 1 + g * (dt*dt) * (d(i−1, t) + 2*d(i, t) + d(i+1, t))/(2*dx*dx), for (0 < i < n−1), and

en−1 = 1 + g * (dt*dt) * (d(n−2, t) + d(n−1, t))/(2*dx*dx), and

fi = −g*(dt*dt)*(d(i, t) + d(i+1, t))/(2*dx*dx)

13 After calculating the new values of h(x, t), correct new h(x, t) to keep the volume of the fluid constant between two steps.
14 If for any index i the value of h(i, t) < b(i, 0) set hi(j) = h(i, j−1) = b(i, 0)-eps, where eps is a small epsilon.
15 Visualize the system using h(x) and b(x).

Using this algorithm and Prompt 55, which was created based on it and completed with details of the user interface and visualization, a program was generated that simulates waves in shallow water. Its results are presented in Figure 55.1.

Task 1. Please, implement the same model in 3D. The paper (Kass and Miller 1990) discusses the technique of dividing the problem into two separate directions. Implement it and visualize the tsunami wave on the beach. Assume that the depth of the ocean decreases until 0, where the wave begins. At time t = 0, assume a longitudinal wave pulse generating a wave with velocity perpendicular to the beach line.

Task 2. Implement a shallow water model based directly on the simplified Navier-Stokes equations from the work of Chen and da Vitoria Lobo (1995) and compare the wave propagation results to those of the Kass and Miller (1990) model.

Task 3. Complete the Chen and da Vitoria Lobo (1995) model to account for the topography of the terrain and the modified wave speed.

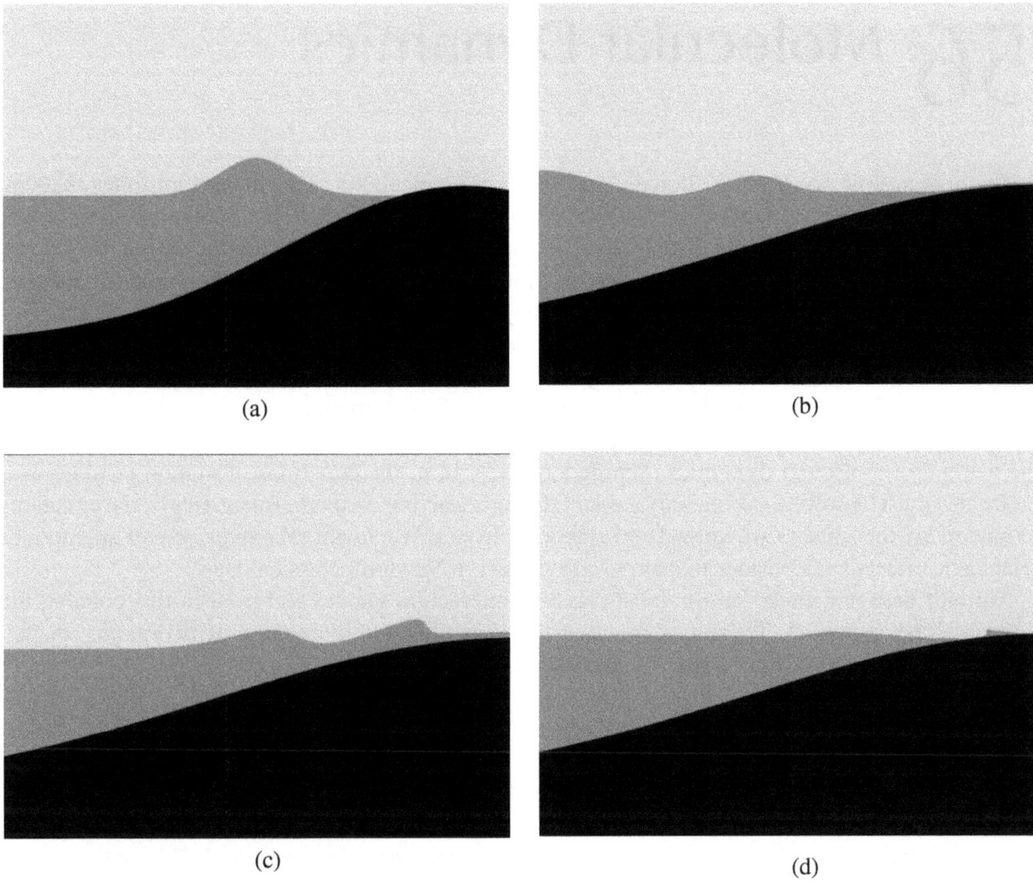

FIGURE 55.1 Wave behavior on the surface of a shallow reservoir at width = 200 and height = 1360, dt = 0.02, dx = 1.0, tau = 0, b(x) = H/8 + H/2.5 * Math.exp(-Math.pow((x – n/1.0)/(n/1.2), 2)), h(x, 0) = H/2 + H/10 * Math.exp(-Math.pow((x – n/4)/(n/10), 2)), and h(x, 1) = H/2 + H/10.1 * Math.exp(-Math.pow((x – n/4)/(n/10), 2)). As the topography of the area changes, the behavior of the wave changes. In the case of shallow water, the wave does not dissolve and lose amplitude but rises and takes on a characteristic asymmetrical shape.

56 Molecular Dynamics

At the microscopic level, gases, liquids, and solids are collections of atoms or molecules that can be more or less strongly associated with each other through interactions. Sometimes (gases) are dilute systems, where individual molecules are not correlated with the motion of others. On the other hand, there are systems where molecules interact strongly, are clumped together in clusters, and arrange themselves so closely together that they form stable structures in characteristic arrangements of crystal lattices. Somewhere in between these two forms are liquids, whose molecules form specific structures. These structures modify as the molecules move, are subject to forces, and transform dynamically (hence, liquids are sometimes said to be soft matter).

To trace the behavior of gases, liquids, and solids at the molecular (atomic) level, one can use the method of molecular dynamics (Schroeder 2015). In it, the equations of motion of individual molecules must be solved, taking into account the forces acting on them. Interestingly, the equations of motion are the same as we know from classical physics. We do not take into account quantum or relativistic effects and consider that atoms are subject to Newton's classical laws.

We will base our model on the most classical interaction known and used in this context, the Lennard-Jones potential. There is a function that describes the interaction of two atoms, distant from each other by a distance r (the "^" sign indicates a power):

V(r)=4*ε*[(σ/r)^12 – (σ/r)^6],

where the constants ε and σ determine the shape of the potential and the nature of the forces acting. It is easy to see that this function goes to infinity for r = 0 (the situation when the atoms approach each other) and to zero when r goes to infinity (the disappearance of interactions). Taking the derivative will show that at a distance r = 2^(1/6)σ, our atoms do not interact. This is the equilibrium point.

Our simulation will be based on the numerical solution of the equations of motion, where time integration alternates with the calculation of forces needed to determine accelerations. We will use the so-called Verlet velocity method, in which we perform several steps with a finite but small dt = 0.01 (we choose the value so that the simulation is stable and conserves energy):

We calculate the new positions, ri, of all atoms using their velocities, vi, and accelerations, ai, from the formula:

ri = ri + vi*dt + (1/2)*ai*(dt*dt),

where the accelerations ai are calculated directly from Newton's second law of dynamics. The forces we get from the gradient of the Lennard-Jones potential after pairs at a specific radius (the restriction on the radius avoids too complex calculations after all neighbors).

We calculate the new velocities:

vi = vi + (1/2)*ai*dt

With the new positions determined, we can count the current accelerations once again (just as we did in step 1).

After calculating the new accelerations, we complete the calculations for the velocities at this time step:

vi = vi + (1/2) * ai * dt.

DOI: 10.1201/9781003734062-57

An important aspect of molecular dynamics is to be aware of the physical context in which the simulation is performed. We will work in a system with constant energy, a constant number of molecules, and a constant volume. Of course, we will allow ourselves to add and remove molecules manually, or to scale the kinetic energy to vary the system's temperature. However, in terms of the simulation run itself, these quantities will be preserved.

The final piece of our puzzle is the initial condition. Here, we will use our knowledge of gas kinematics and assume that, regardless of what state of matter we are in, our particles start from an equilibrium Maxwell-Boltzmann distribution, which will be further supplemented by a procedure for removing macroscopic momentum from the system.

The velocity draw itself will consist of taking a random variable from a normal distribution for each velocity component separately. Determining the variable from the normal distribution is not trivial, and you can use either a library of random numbers for this, or implement an algorithm yourself (e.g., the Box-Muller transformation for drawing uncorrelated pairs from the normal distribution). It is worth mentioning at this point that we will write our simulation, along with the parameters and initial values, in so-called Reduced Variables, where σ = ε = m = kb = 1. This is a standard technique in computer physics, and our solution can later be converted back to variables with physical dimensions.

So let us write one of the most basic molecular dynamics algorithms in a fixed number of atoms, energy, and volume of the system with Verlet velocity integration and the details described above:

1 Arrange N classical atoms.
2 The interaction potential between two atoms:

V(r) = 4*ε*[(σ/r)^12 – (σ/r)^6]

3 Use reduced units: σ = ε = m = kb = 1.
4 The size of the rectangle in which the simulation takes place is 20 × 20. Use rigid boundary conditions (the atom cannot penetrate the walls of the system).
5 Use Verlet's velocity method to integrate the equations of motion of atoms. For each i-th atom, calculate:

New positions: ri = ri + vi*dt + (1/2)*ai*(dt*dt),

New velocities: vi = vi + (1/2)*ai*dt,

Forces are calculated from the Lennard-Jones potential gradient (equation above) for all particles adjacent to it within a selected radius rc=3 (the so-called cut-off radius),

New velocities: vi = vi + (1/2) * ai * dt.

6 ri – position vector of the i-th atom, vi – velocity of the i-th atom, dt – integration time step, 7-During the simulation, calculate the total energy of the system from the Lennard-Jones potential and from the velocity (sum of kinetic and potential energy).
8 Draw the initial values of the velocities of the atoms from the Maxwell-Boltzmann distribution for a fixed temperature, which reduces to a normal distribution for the velocity components. e.g., for the x-component, for a fixed temperature T, it will be:

f(vx) = sqrt(m/(2*π*kb*T))*exp(-m*vx*vx/(2*kb*T)).

9 After drawing the velocity, make sure that the total momentum of the system is zero (calculate the velocity of the center of mass and subtract it from any velocity in the system).

10 Monitor the temperature of the system on an ongoing basis by averaging and scaling the kinetic energy of the atoms accordingly. Show the instantaneous temperature as a graph over time.
11 Conduct a simulation in constant energy (E), volume of the system (V), and number of molecules (N).
12 Visualize atoms as points with a visualization radius rw.

Results for this algorithm, implemented with LLM using Prompt 56 are given in Figure 56.1.

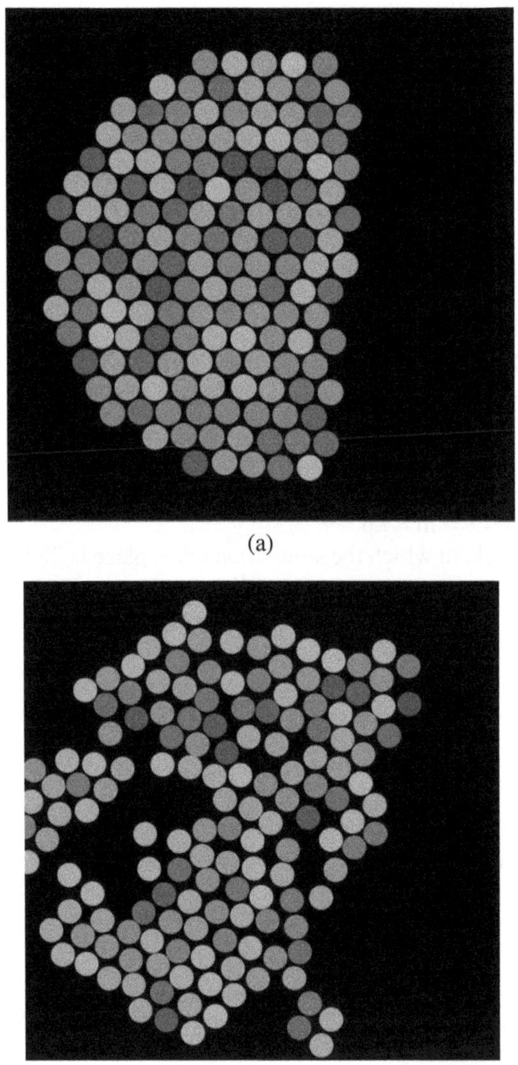

(a)

(b)

FIGURE 56.1 Molecular dynamics simulation of the melting of the cluster and the phase transition from solid to fluid. Results for temperatures T = 0.1, T = 0.432.

Task 1. Draw the Lennard-Jones potential and its gradient to see visually what the forces look like in the molecular dynamics model described in this chapter.

Task 2. In the task at hand, we did not consider the problem of angular momentum, which, in the case of a random initial condition on the velocity, may not be zeroed out and may cause the cluster to rotate. Investigate this effect; determine the rotational speed of the cluster in which the velocities of the atoms were drawn from the Maxwell-Boltzmann distribution as a function of the initial temperature.

Task 3. Propose a solution to the problem from Task 2 without checking the available algorithms in the literature or in the language model.

Task 4. Generalize the simulation to three dimensions and try to generate a tile made of a stable FCC network composed of atoms. Place one free atom on the surface of this tile, let it settle, and find its minimum potential on the surface.

Task 5. Monitor the position of the atom placed on the surface from Task 4 over an extended simulation period. Has the atom changed its position and found another minimum of potential? Or does it make chaotic movements on the surface? Investigate these phenomena using different temperatures of the system.

Task 6. Identify in the literature an interaction potential alternative to the Lennard-Jones potential and conduct a similar simulation with it, addressing the question of what the essence of the impact change proposed in this "other" potential is.

57 Predator Prey Systems

If an ecosystem had only one species living carefree in a vast, food-limited area, it would reproduce at an exponential rate according to the formula (https://en.wikipedia.org/wiki/Lotka – Volterra_equations):

$$dx/dt = ax,$$

where the coefficient a is the rate of this reproduction and x is the number of individuals. This simple equation has little to do with the real world. If our species were, for example, hares, they might occasionally encounter wolves on their path, and the wolves might treat them as food. The reduction of the hare population by being eaten by wolves can be described by supplementing the above equation with an expression proportional to the chance of meeting wolves and hares. Let us denote the wolf population size as y, then:

$$dx/dt = ax - bxy,$$

where b is a constant that determines the rate at which wolves eat encountered hares, and xy determines the chance of meeting them (e.g., no wolves, i.e., $y = 0$, completely removes this expression). A similar equation can be written for wolves. Here, the factors of death in the absence of natural enemies can vary, but in general, the extinction rate depends on the number of wolves present. The expression that describes this will be $-y$, multiplied by the extinction rate. The growth of the wolf population will depend on the availability of food, so there will be a factor xy in the equation, along with a constant that describes the rate of reproduction. The equation describing the wolf population size will therefore be as follows:

$$dy/dt = -cy + exy,$$

where the constants c and e describe the rate of extinction and reproduction in the wolf population, respectively. The pair of equations above is called the Lotka-Volterra model. It describes a predator-prey type system, here with the context being wolf and hare populations. To solve the Lotka-Volterra model, we will use a simple discretization of the derivatives on the left side of the two equations, by which we get the equations in discrete form in the so-called 1st-order Euler Method:

$$x(n) = x(n-1) + (ax(n-1) - bx(n-1)y(n-1)) * dt,$$

$$y(n) = y(n-1) + (-cy(n-1) + e*x(n-1)*y(n-1)) * dt,$$

where the time step number is given in parentheses, and dt is the time step of the numerical solution. After imposing the initial conditions x(0), y(0), and setting the constants a, b, c, and e, it is enough to iterate the above two discrete equations in a loop to follow the evolution of the two populations over time. Such a simple 1st-order differential scheme is not the best choice, of course, because the phenomenon is not linear. There are higher-order predictive-correction methods, which would allow more reliable results for large time steps. However, a sufficiently small step dt allows tracking cycles in the ecosystem, and the method does not introduce noticeable errors into the solution.

DOI: 10.1201/9781003734062-58

So let us write down the Lotka-Volterra model we described earlier:

1 Solve numerically the system of differential equations:

$$dx/dt = ax - bxy,$$
$$dy/dt = -cy + exy,$$

where a, b, c, and e are the parameters of the model.
2 Use the Euler method to solve, where after discretizing the derivatives in time:

$$x(n) = x(n-1) + (ax(n-1) - bx(n-1)y(n-1)) * dt,$$
$$y(n) = y(n-1) + (-cy(n-1) + e*x(n-1)*y(n-1)) * dt,$$

3 Use the following initial conditions and parameters: x(0) = y(0) = 10, a = 1.1, b = 0.4, c = 0.1, e = 0.4, dt = 0.01.
4 Follow the course of x(t) together with y(t) and the function of x(y).

Figure 57.1 depicts the results of the program generated using Prompt 57. We were able to both obtain equilibrium (illustrated by three cycles of extinction and birth of the two populations) and demonstrate that even a simple numerical method like the Euler method can simulate the Lotka-Volterra model at reasonable accuracy with quite stable cycles.

Task 1. Run the simulation for several different values of a, b, c, and e. For various configurations, draw x(y) graphs to see the limits of applicability of the Lotka-Volterra model. Try to find a limit on extinction and reproduction coefficients – what values can we not use, because they lead to the rapid extinction of one of the species?

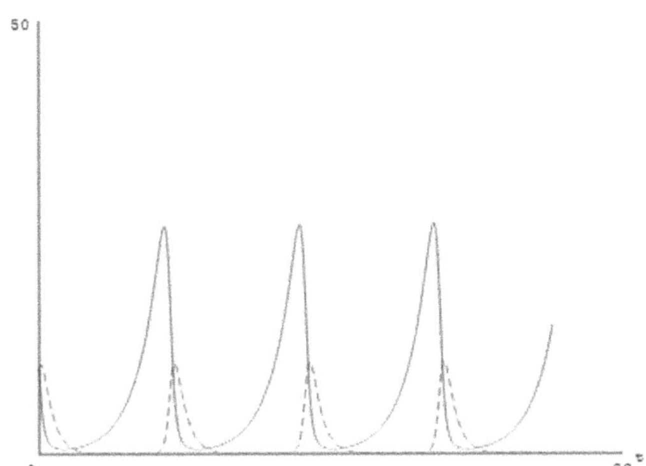

(a)

FIGURE 57.1 (a) dependence of the number of hares (solid line) and wolves (dashed line) on time. (b) dependence of the number of hares on the number of wolves, illustrating invariant and regular cycles in nature. Simulations performed for parameters a = 1.1, b = 0.7, c = 2.7, e = 0.4, dt = 0.001 and initial condition x 0 = 10, y0 = 10.

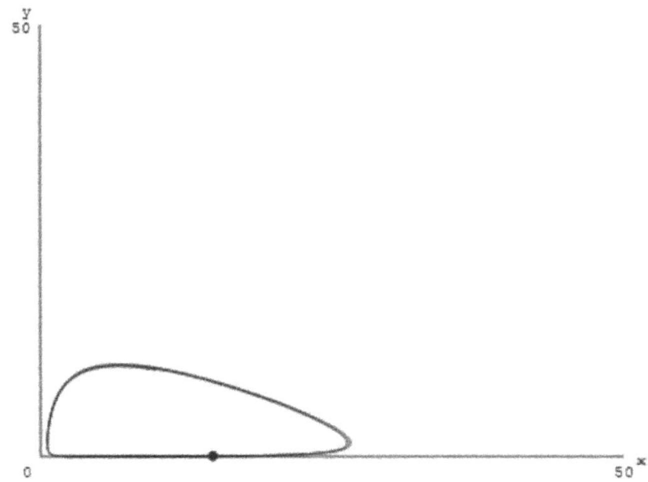

(b)

FIGURE 57.1 (Continued)

Task 2. Generalize the model to three species that form a closed ecosystem. Add a differential equation for the increment z (i.e., dz/dt) and include the effect of this species in the equations for x(t) and y(t). Are you able to obtain sustainable cycles of the three species and their coexistence for a long time?

Task 3. Develop a cellular automaton model, or agent model, where a population x of rabbits on a board reproduces at a rate a and is eaten by a ubiquitous population y of wolves that move around the system. Use the same parameters as in the differential equations we laid out and try to get the same cycles in the population as from the solutions of the above equations. Using visualization, show how extinction and birth cycles are implemented in a model where we can observe individuals.

58 Heat Transfer

Is it worth insulating a building? The answer seems obvious, because insulating will give us energy savings and lower heating bills. However, how to find out whether to insulate with a thick or thin layer of styrofoam? Or is it better to take wool with different properties instead of styrofoam? How does the heat even spread, and how does it penetrate the walls? Will insulating the house help in the summer, when it is warm outside and cold inside the house?

Which place in the house will stay the coolest for the longest time?

Each of the above questions is very interesting because it applies to every home. We will now try to develop a model to answer, at least roughly, each of them. By writing about simplification, I mean reducing the physical processes involved in heat transport to one – diffusion. This is a significant simplification, because at this stage we will not consider convection or thermal radiation.

With many simplifications, such as the assumption of the static nature of heat sources, a homogeneous k-factor in the material under study, we will use the basic heat transport equation:

$$dT/dt = k\Delta T,$$

where we will interpret T(r, t) as temperature, the derivative on the left describes the change in temperature over time, and the Laplacian on the right describes the diffusive process of temperature propagation through time and space. We will interpret the diffusion coefficient k as proportional to the thermal conductivity of the material (https://en.wikipedia.org/wiki/Heat_equation).

We will want to solve the above equation on a two-dimensional projection of the room, where we will reduce the problem to a two-dimensional one and neglect the third dimension (height). We will base our solution on the five-point differential scheme of the Laplace operator, which allows us to approximate the expression of ΔT on the differential grid by calculating the approximate Laplace at point (i, j) directly from the grid values at that point and at neighboring points (https://en.wikipedia.org/wiki/Discrete_Laplace_operator), i.e.,

$$\Delta T(i, j) \approx k * (T(i-1, j, t) + T(i+1, j, t) + T(i, j+1, t) + T(i, j-1, t))/h\text{^}2,$$

where h is the distance between nodes (we assume a square network).

In addition, we will use the linear approximation of the time derivative in the heat transport equation to iterate over time; that is,

$$dT/dt \approx (T(i, j, t+1) - T(i, j, t))/\Delta t,$$

where Δt is the time step size. Our simulation will therefore boil down to iterating the heat transport equation in discrete form, where approximations of the time derivatives and the Laplacian reduce the whole problem to a differential scheme:

$$T(i, j, t+1) = T(i, j, t) + k * (T(i-1, j, t) + T(i+1, j, t) + T(i, j+1, t) + T(i, j-1, t))/h\text{^}2) * \Delta t.$$

The equation should be iterated between two copies of the array T. The idea is not to perform an overwrite of the numerical grid T in place, because entering a new value in it will carry the new value to further calculations (the equation we iterate looks at its neighbors). Similarly, as we have done elsewhere in cellular automata, we keep two copies of the T array (let us call them Ta and Tb) and iterate between them; that is, from the values in the Tb array, we count new temperature

DOI: 10.1201/9781003734062-59

values, which we write into the Ta array according to the formula above. In subsequent steps, we do alternate.

So let us write down the basic algorithm that realizes the issue of diffusive heat transport according to the above description:

1 We are working on a layout with a resolution of NX on NY (300 × 200).
2 Create grids: T with temperature, TWK defining the conductivity of the walls.
3 Start with TWK = 1 in the whole area except for a central frame of 2 grids wide by NX/1.5 side length with hollow holes of 4 grids wide at the corners. For this frame, set conductivity = 0.1.
4 Begin by initializing T with a Gaussian distribution at the center of the array.
5 Iterate over time on the equation:

$$T(i, j, t+1)=T(i, j, t)+ TWK(i, j) * (T(i-1, j, t) + T(i+1, j, t) + T(i, j+1, t) + T(i, j-1, t))/h^2) * \Delta t.$$

In iteration, use the technique of two copies of T grids so as not to overwrite the values. Only after passing the entire grid using the above equation, make a copy to the target array.
6 Visualize the temperature distribution over time using the color map.

Figure 58.1 shows the result of the program generated using Prompt 58. Presented is a map of the distribution of velocity in a building where there are two radiators (lower left and upper right corners of the room), but at some point, someone inadvertently opened two windows – one next to the radiator, the other above the radiator itself. You can see the reduction in temperature throughout the room and the unnecessary energy loss associated with heating the home's surroundings.

Task 1. Conduct a simulation and visualization of heat propagation in your room. Try first to simulate the process of heat propagation from windows and doors, and then see for yourself (organoleptically) whether the results agree with reality.

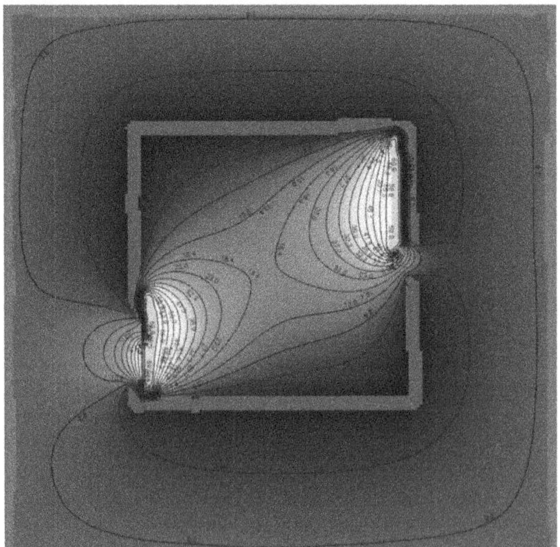

FIGURE 58.1 Temperature distribution inside and outside a house with open windows heated with two radiators at 65 degrees. The temperature outside is 10 degrees, and the walls of the building are impermeable (k 0.000001).

Task 2. Generalize the solution to three dimensions and make an impressive visualization of the surface of constant temperature with transparencies and shadows.

Task 3. Implement the same scheme in the spherical coordinate system. To do this, write the Laplace operator accordingly and write your solution on an irregular grid matched to the problem. As an example, generate a velocity distribution in a spherical-like object with superimposed irregular distortions of different amplitudes, which is heated from the interior and cooled by an object a distance of several radii from its surface.

59 Crowd Evacuation

Who, among us, has not seen in institutions various types of signs, doors, or placards that say "emergency exit"? Placards, or arrows, are needed to direct people in an emergency to places where they can freely leave the place. This applies not only to buildings, schools, or movie theaters but also to vehicles or even airplanes (flight attendants at the beginning of a flight show where the emergency exits are).

Take a group of people in a room where an alarm begins to sound at a specific moment. Each of them starts heading toward the emergency exit. Due to friction with others, the movement may slow down, and the limited capacity of the escape routes will prevent people from exiting.

To simulate this type of simulation, we will describe the evacuation model introduced in the paper (Helbing et al. 2000), where the forces of repulsion, friction, and interaction with walls are taken into account. Models of this type allow us to study various characteristics of such a phenomenon, such as the exit time of all, depending on the speed of the units or the width of the door (Helbing et al. 2002; Matuszak 2010).

As in physics, the motion of each individual is described by an equation of motion. The difference between a physical particle and the person in our simulation is that each person here may have their own internal drive, a will that moves them in a given direction (e.g., toward the door). In the literature, many-particle systems with such a drive are known as self-driven many-particle systems.

Thus, following (Helbing et al. 2000), the motion of each person will be described by the equation of motion:

$$m \cdot dvi/dt = m \cdot (v0 \cdot ei0(t) - vi(t))/\tau i + sum(j, j \neq i)\ fij + sum(W)\ fiW,$$

where the three terms on the right represent forces, the first force matches the speed of the person to the speed indicating the emergency exit door, where vi is the motion vector of the i-th person, v0 is the maximum walking speed, ei0(t) is the unit vector indicating the direction of movement of the i-th person (to the emergency exit), τi is the characteristic time of matching the speed vi(t) of the person to the target speed.

Each person occupies a specific area around themselves, preventing others from getting closer (e.g., a circle with a diameter of 1m). When two people come into contact (the distance is less than the sum of their radii), there is an additional force fij of interaction between the i-th and j-th person between them, which is described by the formula:

$$fij = (Ai \cdot exp((Rij-dij)/Bi) + k \cdot g(Rij-dij)) \cdot nij + \kappa \cdot g(Rij-dij)/(\Delta vtji \cdot tij).$$

The force of interaction between two people consists of repulsion forces (exponential expression), additional repulsion in case of compression (expression with constant k), and the last expression – friction of people between each other (inspired by knowledge of static friction from granular matter). The constants Ai and Bi define the forces of repulsion. Rij is the sum of the radii of the two persons, dij is the difference between their positions, i.e., dij = |ri – rj|, and nij is the normalized vector from person j to person i. Hence, the vector nij = (nij1, nij2) = (ri – rj)/dij. tij is a unit vector perpendicular to nij; that is, tij = (-nij2, nij1). The difference of the perpendicular components of velocity Δvtji is Δvtji = (vj-vi)·tij, and k and κ are constants.

g(x) is a function, taking the value g(x) = 0 for x<0 and g(x) = x for x>0.

DOI: 10.1201/9781003734062-60

The last term of the forces, fiW, we determine for the interaction with the wall. For the distance diW from the wall and the vector niW indicating the direction to the wall and tiW (direction perpendicular to niW), we will use the expression for the force:

$$fiW = (Ai \cdot exp((ri - diW)/Bi) + k \cdot g(ri\text{-}diW)) \cdot niW - \kappa g \cdot ((ri\text{-}diW)/(vi \cdot tiW)) \cdot tiW$$

This movement model will allow us to simulate the evacuation of a crowd from a room and study the exit time based on various parameters.

Let us write down the minimal algorithm for the model we described above:

1 We work on an 800 × 400 board representing an 80 × 40m room.
2 In the center of the room, on a vertical diagonal, place a wall with a door opening 2m wide in the center of the wall.
3 Draw the positions of N people on the left side of the room so that they do not overlap. Each ith person occupies the area of a circle of radius Ri=1m.
4 The equation of motion describes the motion of each person:

$$m \cdot dvi/dt = m \cdot (v0 \cdot ei0(t)\text{-}vi(t))/\tau i + sum(j, j \neq i)\ fij + sum(W)\ fiW,$$

where vi is the motion vector of the i-th person, v0 is the maximum walking speed, ei0(t) is the unit vector indicating the direction of the i-th person's movement (to the emergency exit), and τi is the characteristic time of matching the person's vi(t) speed to the target speed. The formula describes the forces of interaction between the i-th and j-th person:

$$fij = (Ai \cdot exp((Rij\text{-}dij)/Bi) + k \cdot g(Rij\text{-}dij)) \cdot nij + \kappa \cdot g(Rij\text{-}dij)/(\Delta vtji \cdot tij).$$

where Ai, Bi are constants, Rij is the sum of the radii of the two persons, dij is the difference between the positions of the two persons dij = |ri – rj|, nij is the normalized vector from person j to person i, that is, the vector nij = (nij1, nij2)=(ri – rj)/dij, tij is a unit vector perpendicular to nij, that is, tij = (-nij2, nij1), the difference of the perpendicular components of velocity Δvtji is Δvtji = (vj-vi)·tij, and k and κ are constants. g(x) is a function, taking the value g(x) = 0 for x<0 and g(x)=x for x>0.

The last term of the forces, fiW, we determine for the interaction with the wall. For the distance diW from the wall and the vector niW indicating the direction to the wall and tiW (direction perpendicular to niW), we have the formula:

$$fiW = (Ai \cdot exp((ri - diW)/Bi) + k \cdot g(ri\text{-}diW)) \cdot niW - \kappa g \cdot ((ri\text{-}diW)/(vi \cdot tiW)) \cdot tiW$$

5 Use initial values of constants: mi = 80, v0 = 1, τi = 0.5, Ai = 2*10^3, Bi = 0.08, k = 1.2*10^5, κ = 2.4*10^5.
6 When the user presses "Alarm," forces are activated to propel all people toward the exit (door).

Figure 59.1 depicts the results of a program created using prompt 59. You can clearly see the blockage of people trying (most likely in panic) to escape through the narrow doorway. It might be a good idea to widen the emergency exit, but we should either limit the maximum speed of escaping people. As it turns out, the high speed here may be slowing down the evacuation process by increasing the blockage at the exit.

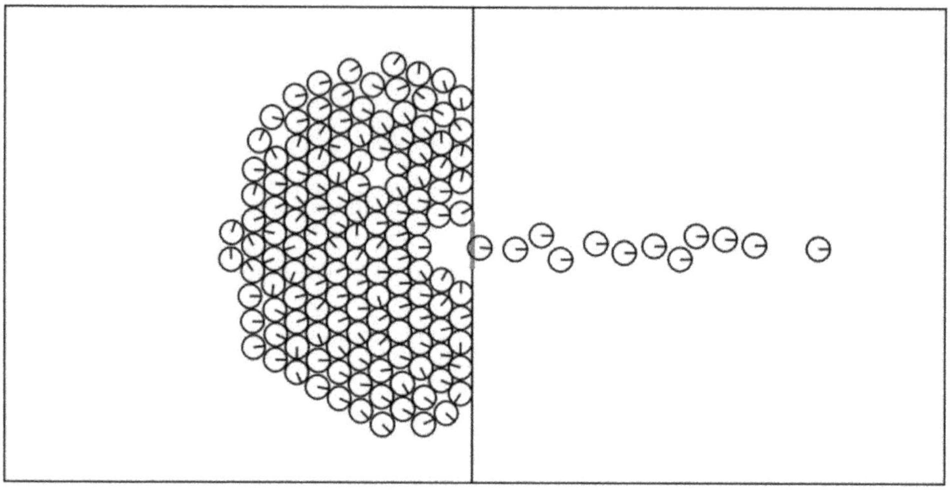

FIGURE 59.1 Evacuation model in action. People in the left room due to the alarm are trying to get out through the door in the middle of the system. Due to excessive speed of movement, friction, and repulsive forces, people were blocked around the door.

Task 1. Add an element of randomness to the model, related to the differences in the behavior of people, the fact that in reality, not everyone follows the crowd unthinkingly; sometimes someone will drop out of a thing, he has to bend down and reach for it, someone will get lost, someone will stumble, etc. To do this, scale the speeds of selected people by a random amount. Investigate how the introduction of such a "temperature" of the system affects the resulting characteristics and evacuation rate.

Task 2. Take into account minor random variations in the radius of the area occupied by each person in the simulation. Check whether such a variation will affect the simulation and evacuation times from the room.

Task 3. Investigate the evacuation time of a room with 300 people in it (counting from the pressing of the alarm to the last person passing through the door). Investigate the dependence of this time on the maximum speed of people in the model (parameter v0). Check under what conditions you can obtain the law of increase in evacuation time with respect to time v0, with a characteristic minimum for small values of v0. See (Helbing et al. 2000).

Task 4. Instead of simulating an equation, simulate the passage of two streams of people through each other. To do this, model two types of people. The first is type A and the second type B. Type A is added with a TA frequency in the upper left corner, with a slight random scatter of positions, and always heads to the evacuation exit in the lower right corner. Type B is added with a frequency of TB in the upper right corner and heads for the emergency exit in the lower left corner. When the person reaches the emergency exit he is heading to, he disappears from the simulation. There are no other differences between persons A and B, except that they are heading in different directions and will be drawn differently. They interact with each other without distinguishing that they are of a different type. For the different parameters of interaction, examine how the flow rates look, how two groups with different interests can exchange, and create a reasonably stable arrangement.

60 SIMPLE (Computational Fluid Dynamics)

The Navier-Stokes flow equations for an incompressible fluid can be written in differential form for momentum:

$$\partial U/\partial t = -\partial U^2/\partial x - \partial UV/\partial^y - \partial P/\partial x + (1/Re)\cdot(\partial^2 U/\partial x^2 + \partial^2 V/\partial y^2),\ \partial V/\partial t$$
$$= -\partial V^2/\partial y - \partial UV/\partial^x - \partial P/\partial y + (1/Re)\cdot(\partial^2 U/\partial x^2 + \partial^2 V/\partial y^2),$$

and continuity:

$$\nabla\cdot VEL = 0,$$

where VEL = (U, V) is the velocity field, P is the pressure, and Re is the dimensionless Reynolds number, which describes the fluid properties, Re = u0*L/visc (u0 characteristic velocity, L characteristic length, visc – kinematic viscosity of the fluid). Solving these equations requires discretizing the differential operators and writing a special algorithm to determine the solution in time. We will now describe a special numerical method that, based on iterations of the guessed flow solution, performs a step-by-step correction of the momentum equations (Patankar 2018). The method is abbreviated SIMPLE, which stands for Semi-implicit method for pressure-linked equations, and in its basic version is used to solve time-independent problems (steady-state solutions). It can, of course, be modified and adapted to non-stationary problems, which we will not do here. In the SIMPLE method, we use pressure to determine the correction associated with the failure of the velocity field to satisfy the continuity equation. For this purpose, we will use Poisson's equation with a source term in the form of deviations from incompressibility of the fluid:

$$\nabla^2 P' = (1/dt)(\nabla\cdot VEL),$$

where P' is the pressure correction due to the non-zero divergence of the velocity field (∇·VEL). The reader may be surprised here; after all, we wrote above that ∇·VEL = 0. Yes, in fact, this is our goal: to make the velocity field incompressible (i.e., ∇·VEL = 0), but sometimes this is not the case, especially in the early stages of the search, when we are basically guessing the initial velocity. Once P' is determined, we use it to correct the pressure and recalculate the velocity fields from the momentum equations. First, we perform the pressure correction:

$$P(n+1) = P^*(n) + P',$$

and then, the velocity correction:

$$U(n+1) = U^*(n)+U',\ V(n+1) = V^*(n)+V',$$

where U' and V' are velocities calculated from simplified momentum equations that take into account the pressure correction P', viz:

$$U' = -\Delta t\cdot(\partial P'/\partial x)\ V' = -\Delta t\cdot(\partial P'/\partial y).$$

DOI: 10.1201/9781003734062-61

The above description outlines the complete procedure for the SIMPLE method (Semi-Implicit Method for Pressure-Linked Equations). After supplementing with the necessary information on the computational grid (we will use the so-called decoupled grid, staggered grid), details related to the implementation and numerical methods, including the selection of finite differences to create differential analogs for the derivatives in the Navier-Stokes equations, we can write down the complete algorithm for solving these equations (Patankar 2018; Matyka 2004). The problem we will solve is lid-driven cavity flow – a standard case in computational fluid dynamics, often used to verify numerical codes in this field. Here is the algorithm for the SIMPLE method described:

0-Let us first determine the necessary equations:

∂u/∂t = – ∂u^2/∂x – ∂uv/∂^y – ∂p/∂x + (1/Re)·(∂^2u/∂x^2+∂^2v/∂y^2), Equation (1)

∂v/∂t = – ∂v^2/∂y – ∂uv/∂^x – ∂p/∂y + (1/Re)·(∂^2u/∂x^2+∂^2v/∂y^2), Equation (2)

∇^2P' = (1/dt)(∇·V). Equation (3)

1. Create arrays of floating point numbers P, P*, U, U*, and V, V*. All arrays are in two copies: one for time n and one for time n+1. The P and P* arrays are NX by NY, and the sizes of the velocity arrays are chosen to impose the appropriate boundary conditions (described later) with a staggered grid of nodes.
2. Set initial fields P*(n), U*(n) and V*(n) to zero.
3. Solve equations (1) and (2) for the fields U*(n+1) and V*(n+1).
   Use a discretization on a differential grid with offset nodes holding the values of the velocity components u (on the top and bottom edges) and v (on the left and right edges), and the pressure P (in the center of the system) – a staggered grid. This defines how you should count the derivatives on the grids. Use:
   - forward explicit discretization for the time derivative,
   - second-order central discretization for second derivatives,
   - forward discretization for pressure derivatives.
4. Solve the Poisson equation (3) to get P'.
   As with velocities, use a differential grid for discretization and compute the solution using the Gauss-Seidel iterative method with SOR (successive over-relaxation).
5. Calculate U' and V'.
6. P(n+1) = P*(n) + P'.
7. U(n+1) = U*(n)+U' and V(n+1) = V*(n)+V', where count the component values from pressure corrections and simplified momentum equations:

   U'=-Δt·(∂P'/∂x)

   V'=-Δt·(∂P'/∂y).

8. Use boundary conditions: left, bottom, and right edges do not move (no-slip condition); top cover moves along the x-axis with velocity V (u0, 0), but only cells from 1 to NX-2 (leave corner cells as no-slip).
9. Visualize the fluid.

Figure 60.1 shows the results of a program created using Prompt 60, based on the presented algorithm and supplemented with implementation details, visualization information, and a user interface.

Task 1. Using the SIMPLE method, simulate the flow past an obstacle in a channel at low Reynolds number. In the channel, use the boundary conditions of the superimposed parabolic Poiseuille flow solution at the inlet and free flow at the outlet. The algorithm we use

FIGURE 60.1 Fluid flow in a lid-driven cavity at Re = 10000. Simulation performed with the SIMPLE algorithm, with visualization using the particle smearing technique (advection with alpha and without clearing the canvas) on top of velocity field mapping higher velocity to lighter color. Parameters used: dt=0.0001, dx = dy = 0.01, u0 = 0.5, kinematic viscosity = 0.00005, NXxNY = 100x100.

here only describes stationary issues, so you cannot see dynamically detaching vortices arising behind an obstacle. To do so, read how to modify the model to solve the time-dependent problem. Implement a new algorithm and increase the Reynolds number to simulate non-stationary flow.

Task 2. Generalize the simulation to three dimensions and study how particles flowing in a cubic flow behave for different Reynolds numbers.

61 Sphere Tracing

Light ray tracing is a relatively simple technique for generating three-dimensional scenes that include lighting, shadows, and reflections. We write about simplicity here because, compared to other standard 3D rendering methods (rasterization), ray tracing is more natural for generating scene lighting, shadows, and reflections.

We will narrow our considerations to one specific algorithm: the ray tracing implementation based on the so-called ray marching and its realization using distance functions (Hart 1996) (sphere tracing). This technique has found application, for example, in the computer demoscene, because it is fast and can be directly programmed with so-called shading programs on GPU graphics card processors.

In sphere tracing, the scene is described by distance functions that specify the geometry by asking for the distance to each object in the scene. A distance function is an analytical formula that, for a given point, returns its shortest distance from the object. Such functions can be with or without a sign (the interior of the object is then treated differently), but the general idea is to be able to estimate how far away an object is from us quickly. It is not difficult to guess that the simplest distance functions are for known solids. For example, for a sphere located at the center of the reference system, the distance function for its surface will be as follows:

$f(r) = |r|-R$, where r is the radius of the sphere and R is its radius.

The functions of the individual objects can be combined, whether using the most straightforward minimum operation (the minimum of the results of two distance functions gives the smallest distance from both of them), or using a smooth transition function (the so-called smoothstep function), which allows you to visually "glue" objects together.

To render an image, first of all, you need to generate, one by one, a ray coming out of the camera in the direction of a flat viewport immersed in the three-dimensional world (i.e., in our case, the area on which we will draw, the canvas) and then follow its course as long as it does not encounter an object close by. We generate such rays from the eye (camera) through each pixel of the viewport, behind which the objects in the scene are located. In general, such an algorithm is a form of ray tracing, and, of course, you can look for object intersections in other ways, either analytically or numerically, but we will use the aforementioned walk. Given the direction of our ray (let us call this unit ray dir) and the starting point (a pixel in the viewport), we can take tiny steps, walking toward an object in the scene or flying out to infinity along the ray. In the first case, we will color the pixel to match the object's color. In the second, well, infinity is probably black or the background color we set ourselves.

However, to perform the above walkthrough, we need to take many steps. Being far away from the objects, however, why do we need to take small steps? It would be better to take one big one. Moreover, this is where the idea of sphere tracing (Hart 1996; https://iquilezles.org/articles/raymarchingdf/) comes in, where we use the information contained in the distance function. We can calculate the value of this function at the point we are currently at on the radius (in the first step, this would be the position of our pixel on the viewport) and move freely by that distance, because the distance function and its values make sure that in a sphere with a radius equal to the smallest distance from the objects in the scene there is nothing!

So, the whole algorithm for performing the steps can be schematically written like this:

Step 1: Start from point p (a pixel in the viewport) and go in the eye-p direction (eye is the position of the eye).

Step 2: Perform at most ITER iterations.

DOI: 10.1201/9781003734062-62

Step 3: Calculate d = f(p) – the distance function of our scene.
Step 4: If d < eps, break the loop (we hit the object).
Step 5: Move by distance d in the direction dir, p=p+dir * d.
Step 6: Return to step 2.

Such large steps (step 5) are a considerable time gain in ray tracing, which is why this technique has gained recognition and found its place in real-time animation.

In addition to ray tracing in the scene, we will add two more functionalities to our algorithm. First, it is worthwhile to calculate the lighting at this point. We will use the simplest Lambert illumination: at the surface, we calculate the cosine of the angle between the light position and the surface normal, which can be easily computed from the gradient of the distance function.

The second element that will significantly enhance our scene is soft shadows. With this technique, you can take an additional walk from the point where we finished our walk toward the light source and sum the distances to the objects in the scene along the way. It is not difficult to guess – if the distances are large, the objects are far away, so you can consider that they are unlikely to obscure the light. We use the sum calculated in this way to appropriately scale the intensity of light reaching the point on the surface where our ray stopped.

The algorithm that implements the above sphere tracing, along with the elements necessary to build a simple scene, is written next:

1 We work on a canvas with a resolution of 1280 × 720.
2 In a loop, for each pixel (px, py), perform an iteration of the ray tracing algorithm:
  2.1 Determine the vector p0 from the eye located at (0, 0, 0) to the pixel on the screen at position (px/RX, py/RY, Z0), where RX, RY is the resolution of the canvas and Z0 is the position of the viewport in the z-axis.
  2.2 Let the vector p initially be p = p0 (the starting point).
  2.3 Let the vector dir be equal to p0. Normalize dir (dir = dir/|dir|).
  2.4 Execute an iterative loop (at most ITER times).
    2.4.1 Calculate the distance d = f(p).
    2.4.2 If d<ε, break the iterative loop.
    2.4.3 Move by distance d, i.e., p = p+dir*d.
  2.5 If the loop has executed ITER times, the color of the pixel is black. Go back to step 2 (continue with pixels).
  2.6 If the loop has executed less than ITER times, calculate IL = the scalar product between the normal vector calculated from the gradient of the distance function at point p and the vector indicating the position of the light located at point (lx, ly, lz). Count the light intensity: light=(1.0+0.5*IL)/distlight^2, where distlight is the distance of point p from the light source.
  2.7 Count the iteration as steps 2.2–2.4.3 from point p to the light source (lx, ly, lz) and count the sum of the smallest distances from other objects in the scene when going from p to (lx, ly, lz). This sum will be our shadow value, which, when trimmed to 0–1, multiplies our light. The final expression for light is thus:
  lightfinal = light * shadow, and this will be the color (brightness) value of the pixel.
3 Perform the above algorithm for the distance function f(x), which returns the minimum distance between the objects of the sphere at the point (0, 0, −10) and the surface with orientation (0, 0, 1). Alternatively, add a second mode for merging objects that uses the smoothstep function instead of the minimum, allowing one object to merge and transition smoothly into another.
4 The camera is set up so that the direction (0, 0, 1) points upward, and the camera faces a sphere located at (0, 0, −10); that is, in the visualization, the surface with this normal vector

will be horizontally aligned. With this camera configuration and scene, we should see a ball centered on the surface.

5 Draw the scene described earlier using the above algorithm.

Figure 61.1 shows the result of the above algorithm implemented in the program generated by Prompt 61.

The technique we have presented here is, of course, much larger than that presented in this chapter, because the field has developed over the years, and we have only presented the absolute basics. We present some of the natural extensions in the form of the following tasks for you to perform on your own, and we encourage the reader who is more interested to make their own discoveries and review the literature (Hart 1996; the website: https://iquilezles.org/articles/raymarchingdf/), and be inspired by the solutions posted on it (https://www.shadertoy.com/).

Task 1. When you hit an object in the scene, you calculate the normal vector. You can also reflect the ray from the surface according to the principle that the angle of incidence is equal to the angle of reflection. You can use such a reflected ray to calculate the next walk (you have a starting point and a direction) and find the object that "sees" the surface from which you bounce. Using this technique, render a scene in which objects are mirrored. Assume 1, 2, and 3 consecutive reflections.

Task 2. Implement the discussed sphere tracing using shader programs (shaders). To do this, implement the algorithm in GLSL and run it in a WebGL view in the browser (example implementations of raymarching on the GPU can be found at https://www.shadertoy.com/).

Task 3. Use the modulo operator, which, when applied appropriately to the argument of the distance function, will cause the object to repeat in space and render a scene of an infinite array of cubes.

Task 4. Instead of rendering a 3D scene, prepare distance functions of primitives in two dimensions and create an algorithm that can render a 2D distance function using them.

FIGURE 61.1 Example rendering of a scene made up of several spheres and a cube. The scene includes Lambert lighting and soft shadows. A smoothstep function has been applied between objects in the scene, making them appear glued together.

Using basic mathematical operations, prepare an interesting distance function to represent the image you invented.

Task 5. Make an animation of a scene of a walking cube, which initially rests on one of the walls, then takes a step in which a 90-degree rotation occurs around an axis passing through one of the edges resting on the surface. After the rotation, another edge currently resting on the surface is randomly selected, another rotation is performed, and so on. To achieve a smooth animation, reduce rendering accuracy or use GPU technology and shaders, as discussed in another assignment.

62 Stalagmites Growth

Stalagmites fascinate researchers because, unlike stalactites, they grow in various forms and shapes, and the mechanism of their formation is not at all obvious. Recent scientific reports explain differences in stalagmite structure by the dynamics of the fall and the impact of drops on their surface. It is hypothesized that one of the factors determining the shape of a stalagmite (it can be flat, columnar, or conical) is the distance between the upper surface of the stalagmite and the surface of the ceiling from which mineral-saturated water drips.

In this section, we will develop a toy model of stalagmite growth. Based on general knowledge of the phenomenon, cursory information about the physical mechanisms, and a few guesses and hypotheses about how these objects are formed, we will build a completely new model that contains the most important elements. Our goal here is not to create a model that reproduces all the shapes of stalagmites, nor to confirm a scientific hypothesis (which is why we call the model a toy). We will use a language model to write code for a problem it has certainly never seen, in the form and content we will present here. The result will be a program that simulates stalagmite growth and lets you "play" with its parameters, turning them on and off.

We will start by defining the basic mechanisms. Our stalagmite will initially grow on a flat surface. Water dripping from above will fall as single drops, spreading gently when they hit the surface. As it spreads, the water leaves some of its minerals on the surface (we will not delve into the physical or chemical processes by which this happens – we will assume it occurs at a constant rate over time).

Because a drop interacts with its surroundings as it falls, turbulent air movement around it can cause chaotic, random movements both in the direction of fall and perpendicular to it. To model this phenomenon, we will use a simple random walk, parameterized by the probability p2 of its occurrence and the deviation of the drop's trajectory from a straight line. This will cause the drop, falling from a high ceiling, to disperse more horizontally and fall further from the projection of its starting point onto the surface. This mechanism was discussed, among others, in (Szymczak et al. 2025), and we will also use it here when constructing the rules of our model.

When a drop falls onto the surface of a stalagmite, it will most likely splash, the size of which may also depend on the height of the fall, which may define, for example, the speed of the drop upon impact. It is worth pausing here for a moment. Of course, it would be ideal to model this phenomenon directly. However, it is not the best idea to model everything in the system 1:1, because we would ultimately end up with a molecular dynamics simulation, or even one from first principles. Instead, we will assume that the splash of a liquid drop and the formation of a characteristic crown composed of small droplets (Harlow and Shannon 1967) is an effect that can be approximated by considering the slippage of the drop on the surface of the stalagmite. When a drop hits the surface of a stalagmite, it does not immediately grow. The drop interacts with a thin layer of the stalagmite, which is usually wet (I remember a lecture where I saw a slow-motion video of a drop falling onto a stalagmite splashing sideways due to interaction with a thin layer of liquid covering its surface). After impact, it must be assumed that the drop may slide to one side. Of course, it usually slides down, and what is more, if it falls into a hole, it is unlikely to come out. We will achieve this by adding mechanics in which a drop falling on a stalagmite "senses" its surroundings and, with a certain probability, chooses to move toward the lowest local height of the stalagmite. This will give us the effect of the mass spreading over the surface of the stalagmite.

Another piece of the puzzle in this model will be the stalagmite's growth. It is tempting to assume that a drop falling on a stalagmite causes growth in that spot with a certain probability. However, this is a too-simple model. Let us imagine a flat stalagmite, a column with a very flat upper surface.

 DOI: 10.1201/9781003734062-63

If a drop hits the very center, it is clear that it will spread across the surface and cause growth there. However, if it hits a curved corner, it will most likely bounce off or run down the wall, and due to the short contact time with the surface (high speed), the probability of a reaction causing stalagmite growth at that point will be lower. We will model this phenomenon with the simple assumption that growth depends on the scalar product of the vector normal to the surface at the drop's location and the direction of the drop's fall (gravity). If these vectors are parallel, the scalar product will be 1; if the vectors are perpendicular (e.g., a vertical wall down which the drop quickly flows), the scalar product will be 0 (zero probability). To diversify the parameter space and allow for modeling different growth modes, we will assume that the formula describes the growth model:

$$p3 = a + b*(n \cdot g)^\wedge c,$$

where g = (0,–1) is the gravity vector, n is normal to the surface, and the constants a, b, and c allow the growth dynamics to be modified.

We will now write down the complete algorithm for stalagmite growth, in which a drop falling with probability p1 downward can make random movements with probability p2 in a direction perpendicular to the direction of fall. The growth process and the probability that the minerals contained in the drop will cause the stalagmite to grow will be determined by the above power law. Additionally, we will add the effect of the drop sliding on the surface. We will write our algorithm as a model of the movement of individual, independent drops that generate a stalagmite. The algorithm boils down to the following steps:

1 We are working in array A with dimensions NX by NY.
2 Initially, clear the array (0), then fill the entire row from 1 to NX at height y = 0 with the value 2 (substrate).
3 Set the position of the particle to px = NX/2, py = NY.
4 In the time loop, perform:
  4.1 Movement with probabilities: p1 for movement py = py–1, p2 for horizontal movement (equally possible either px = px+1or px = px–1). If, after drawing a movement, the molecule would land on a field with A different from 0, reverse the movement.
  4.2 Deposition, if there is a substrate under the molecule (value 2), i.e., A(px, py–1) = 2, then with probability p3 set A(px, py) = 2, where the value of p3 is calculated using the formula:

$$p3 = a + b*(n \cdot g)^\wedge c,$$

  (n is a vector normal to the surface, pointing outside the structure, the '*' sign is multiplication, the '·' sign is scalar product, the '^' sign is exponentiation). The vector g = (0,–1), and 'a', 'b', and 'c' are constants. Calculate the normal vector to the surface as the normalized sum of the normal vectors to the unit faces in the approximate staircase located in a circle with a radius of NC grids from (px, py). After depositing the substrate, place the molecule in its initial position.
  4.3 If the molecule did not settle in the previous step, but there is a substrate underneath it, it performs a horizontal movement (sliding) in the direction of the lowest column. To do this, it checks its surroundings within a radius of NCSLIZG to find which side has the lowest-height column, and moves in that direction by one field with probability p4. If the levels on the sides are the same but lower than the height of the column in which the molecule is currently located, it performs a random movement to the right or left.
5 Visualize array A

In Figure 62.1, we present the results of simulations of the growth of three types of stalagmites, generated using a program written using Prompt 62.

(a)

(b)

FIGURE 62.1 Three types of stalagmites obtained using the growth model discussed here: a) flat type, b = −0.099; b) columnar type, b= 0; and c) conical type, b = 0.1. The rest of the parameters are the same for all types: pl = 0.5, p2 = 0.28, p4 = 0.9, a = 0.1, c = 2, NCSLIZG = 8. The color intensity maps the growth time (color figure available online).

(c)

FIGURE 62.1 (Continued)

The conclusions from our research based on the toy model are surprising. It is not the distance from the ceiling from which the drops fall that determines the shape of the stalagmite. In our case, it is one of the parameters of the growth model, specifically a factor that strengthens or weakens the influence of surface orientation. It was the changes in the range from negative to positive that made the difference in our case.

Task 1. Use fitting to examine whether the theory from (Szymczak et al. 2025) applies to growth in our model. To do this, perform numerical adjustments and try to understand whether the physical deposition mechanisms we used in our model correspond to the assumptions in the aforementioned work.

Task 2. Generalize the model to three dimensions using the exact growth mechanisms. Visualize three types of stalagmites in three dimensions.

Task 3. Using the above description, but modifying the assumptions and physical processes involved in this phenomenon, create a model of stalactite growth. Once you have succeeded, try combining the two models and modeling the moment when the stalactite meets the stalagmite.

63 Jetpack – A Computer Game with Physics

Writing a computer game is one of the more challenging tasks from a programming perspective. The game must take care of several elements that must work together perfectly; there is no room for grinding and inaccuracy; the game user must feel the responsiveness of the controls, get points; there must be information in the game about its state, about the player's gains, etc. These general statements and their meaning, of course, depend on the type of game. However, the universal statement here will be that writing a game is not trivial from a programmer's point of view.

In this section, we will utilize our modeling experience to create a game design that incorporates a built-in physical model for controlling the main character. We will focus on creating a simple game inspired by a game I remember from the days of the Commodore Vic-20 – Jetpack.

In the original game, you control an astronaut who has a backpack hooked up, and his task is to launch a rocket by delivering fuel to it. We will do a similar project, using a simple physical model of the motion associated with the astronaut's jetpack launch. We will assume two states of the astronaut: in the first, he is resting on the surface and can only be accelerated upward. The second state, when the astronaut is already up, allows you to start the side engines and accelerate the astronaut in horizontal directions.

To make the game engaging and allow for points, energy loss, and possible defeat or victory, we will add mechanics in which the astronaut's goal is to collect fuel, fill the rocket with it, and then depart.

So let us write down all the elements of the game in a form similar to the algorithms in other chapters. We will skip the visualization details and focus on the mechanics and elements of the game:

1 We work on a W × H canvas (e.g., 1280 × 720).
2 The game can have two modes: START, PLAY, END.
3 In START mode, write that the game titled "Jetpack" will start when you press the spacebar. After pressing it, the game goes into PLAY mode.
4 The next part refers to the PLAY mode. At the height of H/20 from the bottom surface of the layout (bottom in the sense of displayed at the bottom of the canvas as the user sees it) is a horizontal surface.
5 Place an astronaut on the surface. The astronaut has a height of H/20. The astronaut is described by position r and velocity v.
4 Next to the astronaut is a rocket (empty), which is 2.5x the height of the astronaut.
5 The user controls the astronaut. The down arrow button activates the jet engines at the bottom of the backpack, which eject mass in the negative y-direction in the astronaut's reference frame, causing the astronaut to begin floating. The left arrow turns on the motor facing left in the astronaut's reference system, and the right arrow turns on the motor facing right. Activating the motor causes the mass to be ejected in the right direction and the astronaut to move in the opposite direction. The motors on the sides only work if the astronaut is in the air.
5 Add randomly over the surface, distributed small islands. On these islets, fuel appears randomly in time (with WP frequency). If the hero enters the fuel area, it attaches to their backpack (visible in the visualization).
6 The astronaut can leave the fuel at any time by pressing the spacebar. After pressing the spacebar, the fuel slowly begins its descent phase to the surface of the ground or the islet,

DOI: 10.1201/9781003734062-64

doing so from the drop point in a smooth, downward direction until it stops. If the fuel touches the rocket, the fuel level in the rocket increases (from an initial value of 0), which can be seen in the visualization as the successive coloring of the rocket with the orange color of the fuel from below. Additionally, the hero receives a maximum of 100 points for the fuel brought. However, the number of points decreases linearly to 1 over time, starting the countdown from the moment the fuel is collected.

7 If the fuel fills the entire rocket and the astronaut stands on the ground in the rocket area, an animation of the departure into space is performed. The astronaut flies off into space. The animation should last a minimum of 10 seconds, counting down the points and smoothly transitioning to the end of the game and the win message. Then, the user can enter their name and be added to the list of top scores.
8 Visualize the game as a vector graphic.
9 Add a zoom effect. When the gas button (up arrow) is pressed, the camera zooms in on the character very slowly.
10 Add randomly placed asteroids with a specific TA frequency. Like fuel, they are located on the surface side of the astronaut (above the surface). If the hero enters them, he loses 1/10 of his energy.
11 Display at the top of the screen (on the bar) the passing time, the amount of fuel collected, and the energy bar.
12 After losing all energy, the hero dies – display summary information and let you start again by pressing the spacebar.
13 When the game ends for any reason, the game goes into END mode. Give the player the option to enter a name and write down the top 10 scores.

A full description, in the form of a Prompt, was formulated to contain a description of the visualization, deeper details of the operation, and minor details, as prompt 63. Its result is a program written similarly to the previous implementations of the models in this book. The program is a full-fledged game (see Figure 63.1), allowing you to collect fuel, deliver it to the rocket, and finally fly into space. Feel free to play! The author of the book scored 29 points!

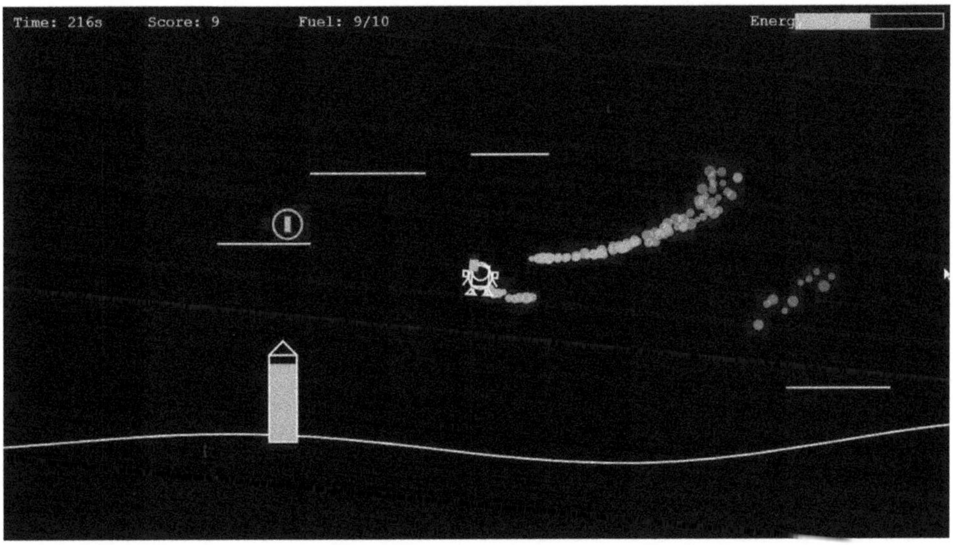

FIGURE 63.1 An excerpt from the Jetpack game created based on the description in this chapter. It contains all the elements described here: a rocket, fuel, an astronaut with control based on physics and the principle of conservation of momentum, asteroids (not visible in the screenshot, but existing in the game), and platforms. The goal of the game is to fill the rocket with fuel (currently 10% filled) and launch it to fly into space.

Task 1. Let an astronaut rest on a planet of radius W/2 (circular, placed in the center of the system) in such a way that the force of gravity always draws him to the center. In such a system, it would be necessary to rotate the astronaut during the flight so that their legs are always facing the planet. Make such a design, leaving the rest of the mechanics unchanged.

Task 2. Find old ROMs of the Jetpack game from the Commodore Vic-20 on the Internet, run them in the emulator, and try to adjust the game so that it is as close as possible to the original game from this computer. Alternatively, you can resort to gameplay posted as videos on the Internet on popular streaming platforms.

64 Chain Fountain (Mould Effect)

In 2013, British science content creator Steve Mould, known for his YouTube channel, presented and popularized in video form the fountain effect, in which a chain initially placed in a glass jar falls out (drops). This would not have been anything special, were it not for the fact that this effect surprised the physics community. To this day, there is no complete and accepted explanation for this phenomenon, and the video has been viewed 2.6 million times (https://www.youtube.com/watch?v=_dQJBBklpQQ). What is more, this phenomenon is so impressive and straightforward that it is often performed, for example, during popular science shows (see Figure 64.1).

To perform this experiment, carefully place the chain in the container, lift it, then grab one end and pull it down along the edge. The structure of the chain and the forces generated in this phenomenon will cause a characteristic arc to form along which the chain will move. Interestingly, this arc can be located above the edge line and looks like the fountain mentioned in the title. The source of this fountain, the bend, and the fact that the phenomenon is long-lasting are unclear (Biggins and Warner 2014), and various phenomena and effects are cited here, for example, the bead selection effect (Biggins 2014), impulses from the container (Gibney 2014), and even the lever effect (Yu et al. 2025). Mainly because of this, there is no single key to modeling this phenomenon, and it is futile to search the Internet for codes that allow for simulation, because apart from https://github.com/mattpalermo/chain-fountain-simulator/, which implements a similar model taking into account springs and pulses transmitted from the container to bounce the beads, nothing else can be found. In this chapter, we propose a simple model of a chain of points suspended on springs, incorporating additional virtual springs to account for the beads' interaction with the container. The points will represent beads, and the springs will have a high enough spring constant that the connections are sufficiently rigid to model the behavior of an actual chain. The main idea here is to have the forces propagate along the chain as quickly as possible.

The chain will initially be placed in the container. To model and control collisions between the chain and the container as simply as possible, we suggest using additional virtual springs that appear when chain points penetrate the edges (walls) of the spherical approximations. The simplest way to do this is to approximate the walls with circles, which, when arranged with sufficient density, will create the wall geometry and be easy to handle from a simulation perspective (since there will be one additional force). In the model, we will neglect interactions between the beads (they will even overlap).

To model the fountain phenomenon, we will allow the user to fully interact with the model and impart an impulse to the string by pulling down one of the beads. The combination of elastic forces on the string, interactions with the container, gravity, and the inertia of the masses that make up our string model will cause a characteristic fountain to form on the chain, which quickly leaves our container.

We will track points by numerically solving their equations of motion. We will address this issue using the Verlet method, which conserves the system's energy (it is important to conserve the chain's energy). Interestingly, because in the literature (or perhaps better said, on the Internet), the term "Verlet simulation" often refers to codes in which, in addition to elastic connections, constraints are imposed on distances, we must explicitly indicate in the algorithm whether we want them or not.

So let us write down the complete algorithm that we will use to create a simulation for Mould's chain fountain effect:

DOI: 10.1201/9781003734062-65

FIGURE 64.1 An experiment demonstrating the ejection of a chain constructed from beads suspended on lightweight connections made of string (own photograph).

Create a two-dimensional simulation in a single JS/HTML file:

1 Create a chain model in a gravitational field.
2 The chain is made of beads connected by Hooke springs with high elasticity and damping.
3 Do not use additional constraints on the distance between points.
4 Make sure that the chain is always completely arranged in the container at the initial moment.
5 The user can grab any point and move it.
6 Design a container without a top cover.
7 Implement a simple collision model with walls (virtual spring).
8 Allow simulation parameters to be changed.
9 Use the Verlet method to integrate the equations of motion of the points.

Figure 64.2 shows the results of the code generated using query 64. The desired effect is noticeable and, best of all, it is created from a straightforward, basic chain model. You can also see the effect of the container's height above the surface. Interestingly, our model also captured the elongation of elastic connections, which indicates much greater stresses arising from a fall from a great height.

Task 1. Examine and draw the relationship between the fountain's height and various parameters: height above the surface, spring parameters, gravity, and bead masses. Find the optimal settings to achieve the highest fountain.

Task 2. Examine the impact of calculation accuracy (numerical integration method and/or integration step time constant) on the resulting fountains.

Task 3. Try to analyze what is happening in the system. Example: color the chain in the visualization to show where the most significant stresses occur. You can also draw the stresses as a function of the distance from the first bead. Develop your own method for analyzing this system.

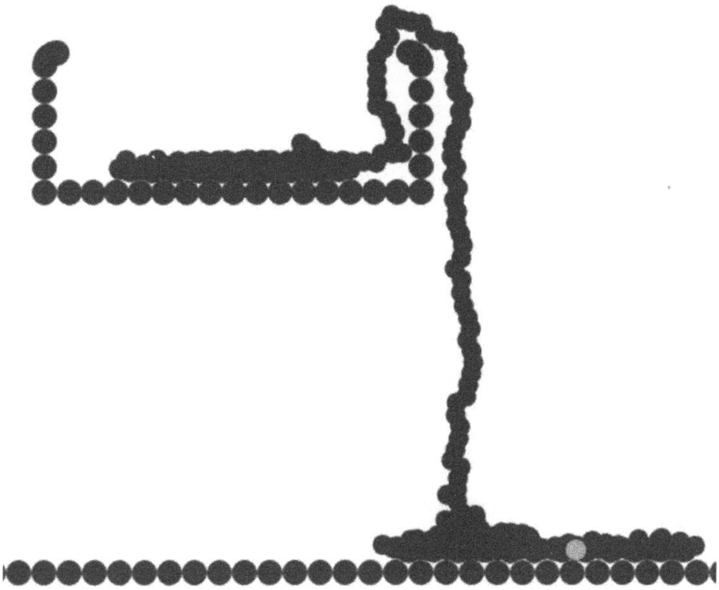

(a)

(b)

FIGURE 64.2 Fountain (Mould effect) from a chain thrown from a container for parameters: N = 2120, R = 8, D = 9, M = 1, g = 1500, k = k2 = 150,000, damping = 0, dt = 0.000125, appropriate box sizes and floor distances: a) d = 300 and b) d = 2500.

Postscript

My vision of compiling as many models as possible, presenting them in the form of natural language prompts, and preparing codes for them using language models led, as I expected, to the point where I have to say: enough. These examples represent most of my ideas and attempts over the course of many years. Some of them I have implemented myself, and others are my long-standing "dreams" to write, run, and see. However, there is a moment in every project when you have to close it, and that is what I am doing now. Next, dear readers, I leave you with a few of my unrealized ideas, inspirations, and projects that could successfully form the basis for the second part of this book in the future:

1. **Rigid body.** A model of a physical pendulum with any density distribution and a procedure for calculating the moment of inertia: writing down the equations of motion, taking into account the moment of inertia of the solid, and solving with 3D visualization. Then, implementation of the collision of a rigid body with a surface and execution of two scenes: a roly-poly object; that is, a properly constructed object with a density distribution, causing it to automatically return to its equilibrium position with the denser part at the bottom, and a scene of 1000 cubes with collisions that pile up into a stable stack and calculation of the angle of repose for such material. As a starting point, it is worthwhile to perform a simple simulation of a rigid body (Hecker 1996). In the prompt, define the basic elements of the scene and the arrangement of objects (initial conditions), describe the moments of inertia, the equations of motion of a rigid body, indicate the numerical method for their solution (Runge-Kutta 4), describe collisions (impulse methods may be a good solution), and indicate the method of visualization and interaction with the user.
2. **Coin rotating on a table.** Simulation of a coin set in motion on a table. A numerically stable simulation method is required for a rigid body, a friction model, and a coin-table collision model. The goal here will be to reproduce the results of Caps et al. (2004), in particular, the 3D visualization of the coin's start, the behavior of the coin's angle of inclination over time, and the calculation of the shape of the trajectory of the coin's center of mass. An interesting development here is the real-time generation and playback of the sound produced by the coin. In the prompt, define the type of simulation (rigid body) and describe its equations of motion, as well as its initial state (position and rotational speed at the beginning). It is essential to select the correct collision model to prevent energy loss during rotational motion (this may be a critical point – what is the mechanism of tilting a rotating coin?). When it comes to collisions, it is worthwhile to implement a model for generating sound by synthesizing sound waves in real time. You can also describe the measurement procedures in a single prompt and define what we mean by "tilt" and "center of mass."
3. **Quantum billiards.** The temporal evolution of a wave packet, initially a Gaussian distribution, through the numerical solution of the time-dependent Schrödinger equation in a two-dimensional potential well geometry with an oval shape. It would be interesting here to compare this to the phenomenon of classical billiards, where particles interact with the walls of a stadium, resulting in classical collisions without energy loss (Christoffel and Brumer 1986). It would also be interesting to compare both classical and quantum phenomena in both stadium geometry (a square with rounded corners) and ellipse-shaped geometry. The prompt should include the numerical method for solving the Schrödinger differential equation, the initial condition, and the method of visualization and collection of results.

DOI: 10.1201/9781003734062-66

4. **1D wave visualization.** There is an interesting way to visualize the propagation of a one-dimensional wave (e.g., a deflection propagating on an elastic string) by drawing successive deflection graphs one below the other in such a way that the time arrow points downwards. The resulting visualizations can be colored, or the amplitude can be artificially increased to obtain interesting visual effects. This technique has been used, for example, to visualize quantum and classical phenomena (Brandt and Dahmen 1989; Matyka 2021). It would be interesting here to clear the wave graph from previous time steps by invisible (empty) spaces under the graph from the new time step (this would create an aesthetic effect and perhaps the illusion of three dimensions). To create the prompt, it is necessary to define the method of numerical solution for the 1D wave equation (or the time-dependent Schrödinger equation) and describe the drawing procedure in detail, as described above.
5. **Hidden visualization of 3D scene.** One of the most interesting ways to visualize functions of two variables, which are the result of many wave phenomena and more, is lines. Creating a grid composed of lines with anti-aliasing enables a reasonably accurate representation of local deformations in the visualized fields and provides an aesthetically pleasing visualization suitable for printing, for example. If we were to draw lines visualizing such a function normally, drawing everything as it appears, for some camera positions, the image would be unbearably confusing because we would see lines in front of and behind each other, one on top of the other. Therefore, it is worth ensuring that invisible lines are removed using a special algorithm. I first saw such visualizations in the book (Baumann 1989), which showed a function of two variables with surface coverage. Coverage here consists of obscuring parts of the function located behind a hypothetical surface. Moreover, this is where the most significant difficulty arises. The removal of such lines can be "tricked" by drawing surfaces parallel to the lines. Although it appears satisfactory, upon closer examination, it is not the optimal solution. With its approximations and high variability of the function, it does not look perfect (we employed this technique in this book in the chapter on 2D waves). The task here will be to construct an accurate algorithm that eliminates invisible lines from the visualization of any 3D scene built from triangles (in particular, a graph of a function of two variables, but not limited to this). The specific goal is to create a scene of a flight over a city built from 3D buildings (simple solids, cuboids) with visualization performed using the algorithm with lines. The prompt must define a specific scene, using an algorithm that determines the lines to be drawn by eliminating the invisible ones. This task can be performed by considering the projection from a 3D scene to a 2D scene and developing your own rendering and rasterization procedures.
6. **Process model (elevator)**. The apartment building where I lived as a child had 10 floors and a ZREMB elevator. On each floor, there was a single button to call the elevator. Interestingly, when we enter an apartment building or hotel today, in 2025, most of them have two buttons on each floor. Why? Well . . . It is more convenient for designers because the elevator does not require programming. My old elevator had one button because its designers thought that a user standing on their floor did not need to know where the elevator was, and that it was not really their job to tell the elevator which way to go. The user would only want to bring the elevator to them, and one button is enough for that. I remember precisely the rules that governed this antediluvian elevator, and they stuck in my memory:

   Rule 1: An elevator going down stops along the way if a button has been pressed at that level.

   Rule 2: The button in the elevator takes priority over the button pressed in the hallway on any floor above it only if the button in the elevator was pressed before the button in the hallway.

   Rule 3: An elevator going up to floor P (called from the starting floor $D < P$ by a button in the hallway or controlled by a button inside the cabin) does not stop at any of the floors $K < P$ along the way, even if a button is pressed there (the elevator remembers its destination floor P and goes straight there).

Rule 1 is still used today by children who, running faster than their parents in the elevator, stop the elevator on subsequent floors. Rule 2 makes sense because we would not want to be surprised after entering the cabin and pressing the button to find out we are going somewhere else. This rule was probably the weakest, as it happened to me that I was going from the 5th floor to the 10th because someone pressed the button faster than I did. This was a consequence of keeping the doors open for a long time and was a punishment for holding up the elevator. The combination of rule 1 and rule 3 is the most interesting. From the point of view of the elevator ride, it is good if it picks up "customers" on the way down but does not do so on the way up. The elevator should go to the top first and then, on its way down, pick up everyone who is standing there. This way, we increase the time during which new "customers" can appear.

Assuming a limit of six people for a ZREMB elevator, we can simulate the process of transporting people in a typical stairwell in a 20-apartment building, where they leave at different frequencies and want to use the elevator. The task here will be to simulate the entire process, including the visualization of queues of people on individual floors, the possibility of changing the frequency of new people arriving, occasional trips to the top, and the definition of simulation properties (number of people, frequency, elevator speed, number of floors, number of apartments, etc.) from the user interface. Finally, I suggest visualizing the elevator with all the queues and measuring the elevator's throughput, taking into account rules 1–3, and then repeating the same process without one of these rules. You can also examine the elevator in a selected building, introduce other rules, and demonstrate the craftsmanship of the one-button elevator engineers, who, I am sure, have worked out these processes very carefully in the past.

7. **Simulation of a ball with a melody.** The idea here is a simple physical simulation of a ball falling in a gravitational field, which, when it hits obstacles in the scene, generates a melody according to the musical notation. The arrangement of obstacles in the scene should be considered so that the melody played corresponds precisely to the musical notation (mainly the time of impact and the associated playback of the sample, as well as maintaining the tempo of the piece). The difficulty here lies mainly in selecting methods for determining the ball's trajectory so that it is physically correct, with correctly calculated angles of reflection, conservation of energy, and no conflicts between the calculated trajectory and other obstacles involved in playing the melody. This complex task can only be accomplished by breaking down the prompt into its constituent parts: sound playback technology, computer simulation, and a physical model of the ball described in such a way that the model can plan the distribution of obstacles. The prompt should specify a procedure algorithm here to create a program that implements an algorithm generating the appropriate arrangement of obstacles, for example, using analytical ball motion trajectories with adaptive (trial-and-error) scene matching to the song being played. The model will not perform such an algorithm for us, but it can save it in the form of a program. The goal is to perform smooth, real-time animation based on this physical simulation, accompanied by the playback of several pieces of music.
8. **Synthesis of guitar string sound.** Physical modeling for sound synthesis is an interesting technique that enables the model to generate and listen to a sound wave resulting from movement, collision, friction, and other physical phenomena. One of the basic models of this type is the Karplus-Strong synthesis, which aims to generate the sound wave of a guitar string. The algorithm is quite simple and boils down to averaging the signal over time in order to dampen it (just as in reality, the amplitude of the vibrations of a plucked string decreases over time). Its specific implementation can be the generation of an initial, periodic sound signal, for example, noise or a square wave, and averaging the signal over time to a new buffer according to the formula (Karplus and Strong 1983; https://en.wikipedia.org/wiki/Karplus-Strong_string_synthesis): Yt = 0.5*(Yt-p + Yt-p-1), where the subscript denotes time and p is the sample size (the delay after which we perform averaging). Such

a simple model produces a surprisingly realistic sound of a plucked string. The prompt should include the string model, specify the technology used to create the buffer for the sound being played, the method of storing the sample and its modifications (averaging) over time, and the method of maintaining the appropriate averaging rate relative to the music being played.

9. **The Lattice-Boltzmann (LBM) method by algorithm.** The algorithmic notation of the LBM method is straightforward and involves determining the primitive vectors of the computational grid, the collision model, and the initial condition. This approach differs from the one we used in two chapters of this book, which dealt with the implementation of flows through porous media. There, we described the algorithm at a higher level of abstraction, assuming that some knowledge is common, which can be done when we are sure of it and can verify the results obtained. However, it is also worth implementing LBM in detail, as this allows for fairly precise adjustments to the algorithm, modifications, and even attempts at extensions such as the moving boundary algorithm, the submerged boundary method, the development of a meshless solver, or the introduction of curved boundaries. The goal here will be to develop a precise prompt that generates LBM code with all low-level elements of the algorithm, without requiring the implementation to fill in the details using existing knowledge in the literature. The result should be a two-dimensional flow with visualization and validation based on a simple flow through a driven cavity in 2D.
10. **I would also recommend considering the following models:**
    - An ocean surface model based on an experimental deflection model (Tessendorf 2001)
    - Drawing arbitrary curves and shapes from FFT analysis (https://www.dynamicmath.xyz/fourier-epicycles/)
    - A study on combining models and combining, for example, pendulum motion with fluid dynamics (generating fluid vortex motion around a physical pendulum constructed from a rigid body) using the technique of mixing prompts or mixing the results of prompts
    - Implementation of a real-time smoke model (Stam 2003)
    - Use of the above smoke model for animation using a dynamic algorithm for inducing vorticity through character animation
    - Spacetime constraints in character movement (jump) simulation (Witkin and Kass 1988)
    - Use of constraints for interactive character simulation in a game (https://www.cs.ubc.ca/~van/sssjava/javademo.html; van de Panne and Lee 2003)
    - Behavior of a flock of birds and signal propagation in a flock (Reynolds 1987)
    - Determination of geodesics in the motion of particles near a black hole
    - Visualization of gravitational lensing (with 3D visualization)
    - Implementation of the LBM model in curved space (Mendoza et al. 2013)
    - Model of wolves and hares on a grid (cellular automaton)
    - Voting model and opinion dynamics (Sznajd-Weron and Sznajd 2000)
    - Anomalous diffusion, particle study with Levy flights
    - Lattice Gas models (FHP, HPP)
    - Multiparticle Collision Dynamics (Matyka 2017)
    - Implementation of an associative network and pattern recognition
    - Visualization of a 3D scalar field using the marching cubes algorithm
    - Moving cube model taking into account friction and internal propulsion of the object
    - Robot arm, kinematic inversion algorithm with user interaction (Lander and GRAPHIC CONTENT 1998)
    - Modeling of river evolution, meandering, and delta dynamics (Cohen et al. 2015; Seybold et al. 2007)
    - Algorithm solving the traveling salesman problem with visualization

- Simulation of processes and queues in supermarkets
- FFT analysis (two-way) of a two-dimensional signal (drawing)
- Implementation of the Position Based Dynamics (PBS) method for rope simulation (Müller et al. 2007)
- Jet aircraft takeoff model taking into account empirical formulas for drag and lift, wing characteristics, and implementation based on the above simple takeoff and landing simulator
- Jet aircraft takeoff implementation taking into account the physical model of the aircraft, wheels, and their interaction with the surface, and a simple model calculating lift based on geometry (Kucharski 2019)
- Implementation of a model of leaves in the wind (Wejchert and Haumann 1991)
- Aikawa attractor (and other attractors) with 3D visualization
- Simulations of slime molds and pathfinding – continuation of the large-scale anthill model, preferably using GPU tools (https://cargocollective.com/sagejenson/physarum)
- Procedural methods of computational fluid dynamics in animation (Bridson et al. 2007)
- Aerodynamic model of a ball shot at a goal (penalty kick) – Magnus force and ball rotation (https://www.grc.nasa.gov/www/k-12/airplane/socforce.html)
- Implementation of the marker-and-cell method to simulate a drop falling into a tank (Welch et al. 1965; Harlow and Shannon 1967)

I encourage you to conduct your own experiments and wish you fun!

Prompts

In that section you will find computer LLM prompts that were developed for this book and used to generate computer codes that created results for all models described in the first sections. Prompts are complete. All we need to do is feed them into the language model of our choice, which should generate the code for us using this one prompt only. You can use them to create your own simulation programs, play with them, extend, and develop further. Be aware, this part of the book was not edited nor corrected as author requested all prompts to be 100% original as those used as the input for LLM's.

Full software, open source, produced using these Prompts are available on the author website and on GitHub: https://github.com/maciejmatyka/Computational-Modelling-with-Single-Prompts-Book

PROMPT 1. MONOPOLY GAME

Write a JS/HTML program that implements a simple model of a 40-element Monopoly game, in which the player starts from field 1 and moves as many fields as they roll on the dice in each step. Set a periodic boundary condition on the last and first spaces. Perform two independent simulations using this model, each time performing an infinite number of trials. In the first version, let the player simply move around the board. In the second version, add a "go to jail" space (space no. 30). If the player lands on this square, they are moved to square 10 (jail). In both variants, count how many times the player landed on each square. Draw the results live during the simulation. At each step, write down the step number and redraw the two graphs, where the x-axis is the square number and the y-axis is the number of times the player landed on that square in the first and second variants. Optimize your solution for speed. Make the plot clear, simple plots black and white, no colors, no extra graphics, with all column visible and not covered by the text and axes.

PROMPT 2. STRIPES ON THE ZEBRA

Write a JS/HTML program in a single file that simulates a cellular automaton of spots according to the algorithm below and presents the results in the form of a black and white map representing the state of the automaton (color = cell state) and its evolution in real time. Add start and reset sliders and the ability to set the probability of the initial state and the grid resolution. Add a graph that shows the number of cells with values of 1 as a function of time. Algorithm:

Step 1. Fill the array with random numbers 0 or 1.
Step 2. Make a copy of the array.
Step 3. For each cell in the array, calculate the sum of its neighbors and add the value from that cell (you will get a value between 0 and 9).
Step 4. Use the rules to select a new value for the cell and write this new value in the copy of the array: 0 -> 0, 1 -> 0, 2 -> 0, 3 -> 0, 4 -> 1, 5 -> 0, 6 -> 1, 7 -> 1, 8 -> 1, 9 -> 1
Step 5. After completing steps 3 and 4 for all elements of the array, copy the copy to the correct array (this can be done element by element).
Step 6. Show the new state of the array and return to step 1. Use periodic boundary conditions in both directions.

PROMPT 3. THE GAME OF LIFE

Write a simple cellular algorithm in a single JS/HTML file with real-time visualization. Let the grid have a resolution of 100x100. Let the algorithm have the following transition rules:

DOI: 10.1201/9781003734062-67

1) A live cell survives only if it has 2 or 3 neighbors.
2) A dead cell will be reborn if it has exactly 3 neighbors.

Algorithm:

Step 1. Fill the array with random numbers (or another initial state).
Step 2. Make a copy of the array.
Step 3. For each cell in the array, calculate the sum of its neighbors (do not add the value of that cell, you will get a value between 0 and 8).
Step 4. Use the above rules to select a new value for the cell and write this new value in the copy of the array.
Step 5. After completing steps 3 and 4 for all elements of the array, copy the array back to the original array (this can be done element by element).
Step 6. Display the new state of the array in black and white and return to step 1.

Allow the user to change the initial configuration to a random state with a random distribution and a specified porosity. Add a small graph showing the dependence of the density of live cells on time. Allow the user to reset the game state and the graph with a button.

PROMPT 4. LARGER THAN LIFE

Write a cellular automata simulation in a single JS/HTML file with real-time visualization on (initially) 100x100 grid. The algorithm:

Step 1. Fill the array A with random numbers 0/1 (dead/alive cells).
Step 2. Make a copy of the array (A->B).
Step 3. For each cell in array A calculate the sum of live cells in its neighborhood (to take sum use the grid B). The neighborhood should also include the cell value in itself.
Step 4. Use the rules mentioned below that determine the new value of each cell in grid A based on the sum calculated in the previous step.
Step 5. Show the new state of the array and return to step 2.

In step 3. When counting the values of cells alive in the neighborhood, consider a wide square or circular neighborhood, where the cell for which the cell algorithm is performed is located at the center of the neighborhood. In step 4, depending on the value determined from the sum in this neighborhood, determine the new cell value according to one of two schemes:

Scheme 1:

R=5 ("bugs" – square neighborhood)

34 <= s <= 58 -> cell survives (1->1)

34 <= s <= 45 -> cell is born (0 or 1 -> 1)

Scheme 2:

R=5 ('stripes' – circular neighborhood)

34 <= s <= 58 -> cell survives (1->1)

34 < s < 45 -> cell is born (0 or 1 -> 1).

For stripes do not include boundary points on the circle when counting the sums. Allow the user to change the variant of the model (either bugs or stripes), the resolution of the grid (by using text boxes where user specifies x and y resolution), and initiate a new configuration to a random state with a random distribution of a given porosity (also given by text by the user). Allow the user to reset the game state and graph using a button. Use a simple two-color visualization (black and white) to display the current state. Don't draw grid lines, just black and white quads representing values in the grid.

PROMPT 5. SMOOTH LIFE

Implement the following cellular automaton algorithm in JS/HTML (single file). Ensure that the code is optimized and reasonably fast (use precomputed indices and omit modulo in computing neighbours in periodic grid):

Step 1. Set RI = 4 and RA = 3*RI.
Step 2. Create an array (e.g., 200x100) of floating-point numbers and fill with random values from 0 to 1 – periodic array.
Step 2. For each cell i, j:
Calculate M – the average value in the circular environment of a cell of radius RI.
Calculate N – the average value in the surroundings of a tire-shaped cell with a smaller radius equal to RI and a larger radius equal to RA.
Calculate the average by taking the total sum of the cell values in the surrounding area, divided by its area.
Step 3. Using the values of M and N, apply the following conditions and determine the new value of s by writing it to a copy of the array:

M<0.5 (dead) => (N > 0.25 && N<0.33)? s(t+1) = s(t) + dt: s(t+1) = s(t) – dt

M>0.5 (alive) => (N >0.38 && N<0.56)? s(t+1) = s(t) + dt: s(t+1) = s(t) – dt

where dt=0.3.

Step 4. After performing steps 2 and 3 for the entire array and creating a full copy of it with the new values, rewrite the entire copy into the array proper.
Step 5. Visualize the grid values by color mapping them to black and white. Then, return to step 2.

Remarks: allow user to control simulation (reset, stop, play) and change resolution of grid and scale of visualization.

PROMPT 6. SIMPLE WAVES

Write a JS/HTML program in one file running on two 400x400 arrays. Use periodic conditions. Write the following cellular algorithm in it:

Step 0. Clear both grids. In array A, place 10 2x2 squares of 1200 at random locations. Copy array A to B.
Step 1. For each cell of grid A, do:

B(x, y) = (A(x–1, y) + A(x+1, y) + A(x, y+1) + A(x, y–1))/2.0 – B(x, y)//wave propagation

B(x, y) = B(x, y) * 0.9//attenuation

Step 2. For each (x, y), calculate:

(a) let v(x, y) = −B(x, y)//velocity.
(b) Calculate the average value:

s(x, y) = (B(x−1, y) + B(x+1, y) + B(x, y+1) + B(x, y−1))/4.0

(c) h(x, y) = s(x, y)*2.0 + v(x, y)//height
(d) n(x, y) = n(x, y) * 0.4//allow the coefficient to change
and show this wave height nicely (height map rendering). Allow the user to change as many parameters of this visualization as possible using the UI.

Step 3. Swap the A and B arrays.
Step 4. Randomly, at a frequency of once per NK steps of the simulation, add a pivot (2x2 as above) at a random location on the grid. Allow the user to select NK from 10 to 400 using the UI.
Step 5. Allow the user to add such square pulses using the mouse (type them into arrays A and B). Allow both – clicking single time to add pulse source or click-and-drag for continuous adding of pulses along moving mouse.
Step 6. Return to step 1.

Visualize height or velocity in grayscale (user choice), allow user to change grid resolution (both x and y), and rescale of canvas. Add the option to turn on and off auto scaling of colors in the simulation.

PROMPT 7. LORENZ BUTTERFLY

1. Write a program in JS/HTML that solves the given system of differential equations and draws an incrementing trajectory (x, y, z) over time in 3D with the ability to move the mouse without other, unnecessary elements:

x’ = s*y – s*x

y’ = −x*z + r*x −y

z’ = x*y – b*z

2. Use parameters: s=10, r=28, b=8.0/3.0 and initial conditions

x = 1.25, y = 1.22, z = 1.2111.

3. Visualize simply with black lines on white background.

PROMPT 8. LANGTON ANT

Write code in JS/HTML that works according to the algorithm shown below. The algorithm works in a 70x70 array with visualization on a 420x420 canvas. The array has two values: 0 (white) and 1 (black). In the program, we will track the ant’s movement.

Step 1. Clear the array.
Step 2. Place the ant in the center of the array and randomly select its direction of movement.
Step 3. The ant moves forward by 1 field.
Step 4. Depending on the color of the square, the ant turns:
90 degrees to the right if the square is white
90 degrees to the left if the square is black

Step 5. The ant changes the color of the square from white to black and vice versa.
Step 6. Draw a board (black and white). Write down the simulation step number.
Step 7. Finish if the ant has reached the edge of the board or return to step 3.

Optimize your solution for speed. Make the plot clear, simple plots black and white, no colors, no grids, no extra graphics.

PROMPT 9. BACTERIAL CLUSTERS (EDEN MODEL)

Write code in JS/HTML that works according to the algorithm shown below.

We are working on a 400x400 array with values of 0 or 1.

Step 1. Clear the array.
Step 2. Place an embryo in the center.
Step 3. Make a list of full cells that are adjacent to empty cells.
Step 4. Select randomly one cell from created list.
Step 5. Select randomly an empty spot from all empty spots around selected cell from previous step. Fill it.
Step 6. Update the list of full cells that are adjacent to the blank cells (make sure there are no duplicates).
Step 7. Show the cluster as a color map (black for 1 and white for 0), also print out the step number and the number of cells added to the cluster.
Step 8. If the number of cells in the cluster is less than 65000, go back to step 4. Otherwise pause the simulation and wait for the user run the simulation further.

Remark: optimize the code without changing the algorithm to parallel.

PROMPT 10. DLA – DIFFUSION-LIMITED AGGREGATE

Write a program in JS/HTML that implements the following algorithm. Work on a 400x400 periodic table initially cleared (value 0). Write a 1 in the array at its center.

Step 1. Calculate the radius R of the circle described by the cluster at the center of the array. To do this, determine the center of gravity of the cluster. Then find the largest circle of radius R that hooks at least one point of the cluster, and for which the rest of the points of the cluster are inside it.
Step 2. Select 20 points at random on a circle of radius R+15 and place them in positions corresponding to the positions of the cells on the grid over which they are located.
Step 3. Each of the drawn points performs a random walk step in one of the four directions of the environment.
Step 4. Suppose any point hits a field adjacent to the value 1 in the array, where it is. In that case, we write 1, recalculate the radius R of the circle described on the cluster according to the above procedure, and draw the position of this point again in a random place on the circle of radius R + 5.

If any point moving randomly goes outside the area of the circle with center at the center and radius R+20, draw its new position on the circle with radius R+5.

On the fly, we display the array of values in the form of a color map on a 2x larger canvas, featuring a white blank area and a black cluster of value 1, along with wandering points. If the cluster has reached the edge of the array, end the simulation.

Show by numerical value the radius of the cluster, the number of attached points, and the simulation step.

Optimize the whole thing (let the animation not slow down the model).

Add a button to add another wanderer at a random location on a circle of radius R+5. Add buttons to select visualization: cluster alone or cluster with wanderers alone.

PROMPT 11. MANDELBROT SET

Write the program in a single HTML/JS file. We work on a 600x500 rectangular image. For each pixel of the image, we choose a color based on the following algorithm:

Step 1. Map the position (px, py)->(x, y) of a point in the image to variables in the range:

px->x, where x belongs to (−2, 0.5)

py->y, where y belongs to (−1.1, 1.1)

Step 2. z=(0, 0)
Step 3. c=(x, y)
Step 4. as long as z.x^2+z.y^2 <= 4 and iteration < MAXITER

{
z = z^2 + c//remember to do this, considering the complex domain
Iteration++
}

Step 5. Color the pixel with the RGB color corresponding to the iteration number (here, you can come up with some interesting color mapping).

While executing the above algorithm, draw the result on the screen in real-time.

PROMPT 12. THE THREE-BODY PROBLEM

Write a JS/HTML program according to the following algorithm:

Step 1. Set the positions of the three bodies and the initial conditions (previous step in Verlet's method) corresponding to the velocities according to the following spread:

x[0] = 0.97000436; y[0] = −0.24308753; x[1] = −0.97000436; y[1] = 0.24308753.

x[2] = 0; y[2] = 0; vx[0] = 0.93240737/2.0.

vy[0] = 0.86473146/2.0; vx[1] = 0.93240737/2.0.

vy[1] = 0.86473146/2.0; vx[2] = −0.93240737; vy[2] = −0.86473146.

Step 2. Perform the integration of the equations of motion of these bodies in a loop:

d^2r1/dt = F12+F13

d^2r2/dt = F21+F23

d^2r3/dt = F31+F33,

where, in general, the Fab forces exerted by body b on body a are: Fab = rab/rab^3, and the vector rab=rb-ra.

Use Verlet's method for integration:

r(n+1) = 2*r(t) – r(t−1) + dt*dt * F(t)/m, where F(t) is the sum of the forces acting on this body (from the above equations). For step zero, calculate the "previous" positions so that the initial velocities from step 1 agree.

Step 3. Visualize using black on white background for drawing points and their full trajectory. In addition to the trajectory and points, draw instantaneous velocity vectors attached to them, scaled so that their length is between 10% and 30% of the average distance between the points. Add a button to reset the simulation state that clears the canvas.

Allow the user to change the time step dt selection in the range of 0.01 to 0.4.

PROMPT 13. RANDOM WALK

Make a program in JS/HTML that implements the following algorithm:

Step 1. Set the position of the particle to 0, 0, 0.
Step 2. Make a random step in 3D space by choosing an angle randomly from a continuous distribution.
Step 3. Store and remember the position of the particle.
Step 4. Make a visualization in which a spline curve is fitted to the positions of the particles. Choose a type of spline such that its shape changes only locally after adding new point to the trajectory. The visualization should be in shades of gray with glued, smoothly curved cylinders representing successive fragments of the trajectory, as well as lighting (at least two sources) and a technique in which the structure casts a shadow on itself.
Step 5. Return to step 2.

PROMPT 14. GALTON BOARD

Make the following simulation in a single HTML/JS:

Step 1. Distribute circular obstacles of radius Robs at a distance of 4Robs in a triangular mesh, so that the first row has one obstacle at the very center, the second row has two obstacles, the third row has three, and so on (total 14 rows visible).
Step 2. Place a ball (radius R) above the first row, randomly shifted +/− 0.1R in x direction.
Step 3. Simulate motion of the ball in a gravitational field. Ball collides with obstacles.
- **Step 3a.** Collisions are realized exactly (ball-obstacle are two circular objects and the exact response to collision may be applied here). Use damping parameter to dissipate kinetic energy during collision. There is no ball-ball collision if many balls are released at once.
- **Step 3b.** Count the number of balls falling on the floor of the system. Use histogram with the number of buckets equal to the number of obstacles in the last row+1.
- **Step 3c.** If ball falls on the floor: remove it from the model and add go to Step 2.
- **Step 3d.** Visualize obstacles as filled and balls as open circles. Use a simple black and white colors. Draw last 100 trajectories of falling balls. Normalize the histogram to a floating-point number between 0 and 1, so that the sum of its values is 1. Draw the histogram under the simulation canvas (whole distribution as columns) such that its total width is the same as the width of the last, most bottom row of obstacles, and the height is rescaled such that the highest column fits the entire vertical space left for the histogram. Print a number of balls for each column. Don't let histogram to cover simulation visualizations and make the whole simulation and histogram visible for the user without scrolling the website.

Allow user to change damping for collisions (in wide range, preferably with slider and text field), the frequency of adding new balls (so that more simulations can take place in parallel), gravity, radius

Robs and R. Add a button to clear the histogram, a button to clear the trajectory, and buttons to stop and resume the simulation.

PROMPT 15. ISING MODEL

Write a program in a single JS/HTML file that simulates the following model: We are working in a 120x80 array with periodic boundary conditions.

Step 1. Fill the array with +1 values.
Step 2. Randomly select one of the cells from the array.
Step 3. Invert the sign of the selected cell and count the induced energy change "DE" (directly from the Hamiltonian H = –J \sum_{i, j} s_i s_j), where s_i is the selected cell and j runs over its nearest four neighbors. To do this, you need to count the energy before and after changing the value in the cell.
Step 4. If DE < 0 -> accept the change. If DE > 0 -> accept the change with probability exp(-BETA * DE). – Repeat steps 2–4 1000 times before visualizing the array and counting the magnetization.
Step 5. Visualize the array as a black-and-white board (five times larger for better visualization).
Step 6. Go back to step 2. Assume J=1, BETA=1/T, where T is the temperature. Allow the user to change T (slider), reset to the initial state with all values set to +1 (button), with the graph cleared and the step set to 0. Live monitor (graph) the magnitude of M as the average of the values from all cells in the array. Print under the graph the current properties, M, step, etc.

Additional task: Add a button to add the current properties to the text box, where the results are stored in the form: T <|M|>. The average of the absolute values of magnetization <|M|> should be taken from 40 previous configurations, 20 MC steps apart each (reminder: 1 MC step = 1000 swap trials) – so you need to remember these values, and the minimum number of steps should be 40*20. Add a "Draw graph" button that draws the values from the text box as a separate graph of T(<|M|>) along with a line-drawn theoretical solution of the Onsager function M=(1−sinh(2.0/T)^(−4.0))^(1.0/8.0).

PROMPT 16. GRAINS IN A SILO

(For 3D version: Add the following sentence in front of the prompt below, "Make this in 3D to pack spheres in cube container:")

Write a simulation in JS/HTML according to the following algorithm:
We are working on a 600x400 canvas.

Step 1. Draw N=25 disks with radius R=W/600.0 (W – width of the canvas).
Step 2. In a loop for each disk:
Step 2.1 The disk increases its diameter by EPS=0.01*W. Step
2.2 Check if a disc overlaps any other: If no, go to Step 3. If yes – a) move both colliding disks apart in a loop (with a step of EPS2=0. 001*W) until there is no collision with another disk – if there is – continue to move, together with the other disk (as if they were glued together) and then if there is further collision – with the next one and the next one (to make sure that no pair ever overlaps) b) apply rigid boundary conditions at the edges (disks do not overlap at the edge when moving, nor when rising). At each step and move make sure not to overlap disks.
Step 3. Continue loop over disks.
Step 4. Visualize the disks in black and white (as black circles, empty and white inside) during the above procedure.

Step 5. Terminate the simulation if you get a configuration where moving grains is no longer possible (checking successive bodies and trying to move them together leads to a situation where the disk would have to move in two directions) – we then have a densely packed situation, return then to the last configuration where no disks overlapped and terminate the simulation. Allow the user to set the number of balls, start, pause, and reset to a new state for the new number of balls.

PROMPT 17. DOUBLE PENDULUM

Write a program in a single HTML/JS file implementing the following algorithm:

Step 1. Initialise the variables: angles and velocities for the double pendulum (4 variables).

Step 2. Integrate the following equations of motion of the double pendulum to track the motion. Use the Runge-Kutta method 4. Equations of motion for both angles:
for angle 1) [1 + sin^2(θ2'-θ1')] θ1" – sin(θ2'-θ1')cos(θ2'-θ1')θ1'^2 + (g/l) sinθ1 = sin(θ2-θ1) * (θ2'^2 + (g/l)cosθ2)
for angle 2) θ2"+(g/l)sinθ2 = –[sin(θ2-θ1)θ1'^2 + cos(θ2-θ1)*θ1"], where θ1, θ2 are the yaw angle of the first and second segments respectively. Denote ' (prim), is the derivative, " is the second derivative. Determine the accelerations symbolically from the equations and use them for integration.

Step 3. Visualise the pendulums as two black and white segments. Next to the pendulum visualisation, draw a graph showing the trajectory in phase space θ2(θ2') (continue drawing until the user clears by clicking 'clear'. Provide the user with the option to change the DT time step (to be written by hand in text field), pause, reset, and set the start angles.

Step 4. In a separate canvas draw the phase diagram of the second pendulum, i.e., trajectory θ2(θ2'), where on the x-axis is the inclination angle of the second pendulum segment and on the y-axis is its angular velocity.

Use simplistic black and white visualization with thick black lines on white background. Optimize for speed. Rescale the phase diagram dynamically if the pendulum goes beyond single loop (when making full rotations).

PROMPT 18. CHAOS IN DOUBLE PENDULUMS

Write a program in a single HTML/JS file implementing the following algorithm:

Step 1. Create angle and angular acceleration variables for 100 double pendulums of equal masses and lengths, in which a user-specified angle alpha1 and alpha2 deflect each. However, a random number in the range 0.0001*alpha1 is added to alpha1.

Step 2. Integrate the following equations of motion of the double pendulum to track the motion. Use Runge-Kutta method 4 (RK4) and solve the equations of motion for both angles for each pendulum separately:
for angle 1) [1 + sin^2(θ2'-θ1')] θ1" – sin(θ2'-θ1')cos(θ2'-θ1')θ1'^2 + (g/l) sinθ1 = sin(θ2-θ1) * (θ2'^2 + (g/l)cosθ2)
for angle 2) θ2"+(g/l)sinθ2 = –[sin(θ2-θ1)θ1'^2 + cos(θ2-θ1)*θ1"], where θ1, θ2 are the yaw angle of the first and second segments respectively. Denote ' (prim), is the derivative, " is the second derivative, '^' is the triplication, g is the gravity constant and l is the length of the segments of the two pendulums.
Determine the accelerations symbolically from the equations and use them to solve using the RK4 method. Make sure the solution is correct.

Step 3. Visualise all the pendulums as two black lines on a white background.

During the simulation, calculate the distance in phase space between the first two pendulums. Monitor (graph) this distance continuously throughout, label the axes on the graph, graph in black and white (black dashes and lettering on white background). Determine the Lyapunov exponent on an ongoing basis. Write out the theory – how you calculate the exponent and how you define the distance between two pendulums in phase space. Allow the user to reset the simulation, set the lengths of the pendulums, the principal angles, and the integration step dt.

PROMPT 19. LOGISTIC EQUATION

Write a program in a single JS/HTML file according to the algorithm:

Step 1. In a loop, run through the values of r from 0 to 4.0 (allow the user to select the range of the loop after r, the step of this loop, reset and start the calculation from the beginning) and execute:

Step 1.2. In an inner loop, execute 1000 iterations of x = r*(1 – x), starting at x=0.5.

Step 1.3. After finishing the above loop, perform next 10 iterations and draw values of x on the graph of x(r).

Step 2. Update the graph and make another run of the loop over r. Let the graph show all the calculated final values of x.

Use black and white colors for the plot. Annotate axes. Allow the user to select x and y ranges and when these ranges are selected, redraw all calculated data (so remember them).

PROMPT 20. SELF-AVOIDING RANDOM WALK

Write a JS/HTML program in a single file:

We work on an 800x600 canvas and perform a random walk without self-intersections:

Step 1. Check if the wanderer has at least one field around him that he has not been to yet. In a situation where the wanderer has blocked, and none of the four fields are available for the walk. Remember the trajectory and start a new walk.

Step 2. Perform the random wandering step by randomly selecting one of the available directions.

Step 3. Visualize all trajectories as black paths (lines alone, no points). The lines should come to the edge of the layout (box).

Step 4. Allow the user to start, reset, stop the simulation, and set the grid resolution (initially 80x60).

PROMPT 21. SNOWFLAKE

Write a program in JS/HTML according to the following algorithm:

1- Start in a 120x120 array with a visualization on a 600x600 canvas.
2- Start with a blank array and a germ in the middle (value 1).
3- Define the neighborhood so that the even rows are shifted by half a cell to the right of the odd rows, i.e., our square array represents a triangular network.
4- Any empty cell that has the number of full neighbors from a given set of values, becomes full (dissolving). Allow the user to enter from 1 to 5 values of the number of full neighbors for which the cell becomes full (so that the user decides these rules).
5- Draw the array on the canvas (black color value 1, white color value 0) and continue the simulation. Remember to move the even rows when visualizing the triangular network.

6- Allow the user to choose the number of steps to execute, reset to the initial state, and choose the initial state (1 cell, two, three, and other different layouts).
7- Allow the user to change the resolution of the array and canvas.

PROMPT 22. CHAOS IN DOUBLE NUMBERS

Write a program in JS/HTML:

1- a=0.434 and b=0.43400001.
2- In a loop (from 1 to 1000), multiply each number in each step by 1.999.
3- If any of the numbers is > 1, then subtract one from it.
4- 4-Draw the scattered point plot of a(b) with a on x axis and corresponding b on y axis (simple style, black on white background).

PROMPT 23. MONTE CARLO Π ESTIMATION

Write a program in JS/HTML:

0- n1=0.
1- Draw in succession N=10000 pairs of random numbers (x, y) in the range 0–1.
2- For each pair, check that the point (x, y) lies in a circle of radius R=0.5 with center at the point (0.5, 0.5). If so, add 1 to n1.
3- Visualize the points on the fly by drawing:
   a) a black square with vertices (0, 0), (0, 1), (1, 0), and (1, 1)
   b) the circle inside this square
   c) drawn points in black outside and green inside the circle
   d) next to it a graph showing a live number: 4*n1/N (over time, all values)
   e) allow user to adjust maximum N and rescale the graph in time accordingly to cover the whole range of steps.

PROMPT 24. MAXWELL-BOLTZMANN DISTRIBUTION IN GASES

Write a program in a single JS/HTML file according to the algorithm:

1- Distribute the positions and random velocities of N gas molecules as rigid 2D disks of radius R, located in a square of side L. Let R=L/100.
2- For each molecule, calculate the displacement vector at the next time step.
3- Check whether the displacement of any pair of particles will cause them to collide.
4- If so, calculate the time of collision (the exact moment of their contact) and do this for each pair of particles. For each such event, memorize the DT step for which the contact occurs.
5- Sort the array of memorized DT steps.
6- Perform an offset for the smallest of the DT steps, which is the moment of contacting specific disks, and for these disks, perform the elastic collision procedure after this offset
7- After recalculating the two-body elastic reflection accordingly, integrate further, taking into account the next DT step from the sorted list to perform the solution of the collision problem of consecutive pairs each time.
8- visualize the system as a simple black and white drawing of disks in a box in real time
9- After solving all collisions and performing the entire time step, return to step 2
10- live draw a graph (histogram) of the absolute value of the velocity and two additional graphs of the velocity histogram, vx and vy
11. Allow the user to change N, L, reset, pause, and start and properties (number of bins) of histograms

PROMPT 25. FOREST FIRE (PERCOLATION)

Write a program in JS/HTML according to the algorithm:
Create an array NxN with values 0. N1=0, N2=0.

1- Perform a loop on c from value 0.01 to 0.99 with step dc=0.01. For each value of c, perform p repetitions to take averages to graph.
2- Fill the array with one value randomly, with a concentration of c. Calculate the number of ones (N1).
3- Insert the coordinates of all cells in the first column into the queue
4- Perform the following steps:
    4.1- Remove the oldest added entry (coordinates) from the queue and in the grid, and in this place, enter the value 2. Add N2=N2+1 (number of lit trees).
    4.2- Execute a loop over the four neighbors of the cell whose coordinates have been removed from the queue (for edge cells, take only the available neighbors). If any neighbor has a value of 1, add its coordinates to the queue.
    4.3- If there are any elements still in the queue, go back to 4.1.
5- Draw the grid configuration on a 600x600 canvas by drawing grids in green for 1, red for two, and white for 0. Side draw live three graphs <N1>(c), <N2>(c), and <(N2/N1)>(c), where < > means the average after K different realizations for the same concentration c.

Allow the user to start, pause, reset, and change the grid size N and p-quantities of repetitions to averages, as well as step dc.

PROMPT 26. GRAVITATIONAL POTENTIAL

Write a program in JS/HTML:

1- Work on a 1000x600 grid, floating decimal values.
2- Assume that there are 3 point masses in the grid (draw their positions and masses)
3- In each cell of the above grid, determine ri (the distance between the center of the cell and mass i) and count: Pi = −G*M/ri, where G=1, M=1
4- Calculate the sum of Pi in that cell (contribution from each of the drawn masses) and record it in that cell of the grid.
5- Draw a scalar map (shades of gray) so that the highest value in the array is white and the lowest is black, and show the masses as small black dots
6- let the user calibrate the visualization by selecting the gray shade for the minimum and for the maximum value in the array, specify the min and max potential for mapping the color to 0–1, multiplier for the array value, multiplier for the color, multiplier for the position of the points.
7- let the user add and subtract weights (mouse), set their values, and reset the whole thing
8- plot only gravity potential field in the canvas (no labels, massess etc.)

PROMPT 27. SAND CELLULAR AUTOMATA

Write a JS/HTML program:

1- Consider a 200x100 grid, values 0 and 1, conditions zeroing values on edge cells
2- fill the layout with random values of 0 or 1 with probability 0.5
3- Execute an algorithm that in 2x2 grids passing through the entire layout for steps

even starts from cell 0, 0, and in odd steps from cell 1, 1, performs the following value swapping if the 2x2 pattern agrees step t, swapping to the configuration at time t+1 (written here as the 2x2 value at time t and next to the 2x2 value at the next time step):

Rule 1

10 00
00 10

The first rule states that the seed located in the upper left corner of a 2x2 grid, with available space underneath, will drop one grid down in the next time step. The following seven situations are already easy to understand by analogy:

Rule 2

01 00
00 01

Rule 3

10 00
10 11

Rule 4

10 00
01 11

Rule 5

01 00
10 11

Rule 6

01 00
01 11

Rule 7

11 10
10 11

Rule 8

11 01
01 11

The final, ninth rule will deal with the situation when two grains, glued together, fall together. In such a situation, in order to model something like friction, or resistance, we will let such grains fall with a certain probability "p", which will change the nature of sand falling at high density:

Rule 9

11 00 (with probability p)
00 11

4- the bottom surface should have a free-flow condition, that is, the values in them are cleaned in each step.
5- visualize the values 0, 1 from the array on a 600x300 canvas, the values one black, the values zero white, and continue to execute the above algorithm
6- allow the user to clear all boundaries (by pressing button) and add boundaries by clicking right and dragging the mouse to draw a line of boundary cells (boundary cells block falling sand).
7- allow user to add continuous source of sand by clicking on the canvas
8- optimize for speed

PROMPT 28. SIERPIŃSKI TRIANGLE

Write a program in JS/HTML:

1- Determine a triangle with side a
2- Draw a vertex inside the triangle (or pentagon, hexagon etc. – to be specified by the user).
3- Draw one of the vertices.
4- Move half the distance towards the drawn vertex.
5- Draw a vertex and return to point 2.

Initially, draw on an 800x800 canvas. Visualize only the points put in the algorithm (black and white drawing). Allow the user to choose the side of the triangle "a", the resolution of the canvas, and the number of points to be drawn, start, pause, etc.

PROMPT 29. DENSE PACKINGS

Write a program in JS/HTML:

0- We start with n=1 iteration.
1- We work on a square canvas of size L.
2- We define a function g(i)=1/i^b, where the sign "^" is raising to a power, i is the iteration number, and b is the given exponent.
2- For the given exponent "b", choose R0 so that g(0)*R0<L/6.
3- Increase the iteration count n=n+1. Calculate R(i) according to the function R(i) = R0 * g(i), where i is the iteration number.
4- Randomly find a place in the layout to insert a circle, star, randomly rotated quad or a tire shaped donut of size R(n) so that it neither overlaps another existing object, do not touch it and is not placed inside of it. The only exception is tire (bearing) that is empty inside. If you cannot find the place in K tries, go to step 3. Make sure, that rotated objects are not touching each other (take into account the rotation while checking the collisions).
5- Visualize all objects (each of different color).
6- Go back to point 3. Continue until the user stops the program.
7- Allow the user to change L, initial R0, allow R0 conversion, change K, exponent b, reset (clear), pause.
8- Allow the user to change the mode to rotated squares, stars, or bearings (tire) with a given inner and outer radius (to be changed by the user), then let the packed figures be added with a random angle in the range of 0–360 degrees. Take this into account when checking if a figure overlaps with another figure that is already placed in the box. Apply periodic boundary conditions to figures passing through walls and take into account passing objects in collisions. Visualize all placed figures as filled figures in random colors.
9- All parameters may be input via sliders and must be possible to change them also from textbox (for wide range).

PROMPT 30. LISSAJOUS FIGURES

Write a program in JS/HTML:

1- The solution to the motion of the pendulum is determined by the formula for the deflection of the pendulum in time: f(t)=f0sin(w0t+fi), where f0 is the amplitude of the oscillation, w0 is its frequency, and fi determines the initial phase of the oscillation.
2- Determine independently the deflection of the three-dimensional pendulum in time separately for the x-axis and y-axis using the formula from step 1.
3- For the given initial conditions and f0, w0, and fi for both directions of pendulum deflection, calculate the position of the pendulum at time t.
4- Draw a point at the determined position on the plane.
5- Continue the loop in time by drawing successive pendulum positions (connect the current point where the pendulum is with the previous one).
6- Allow the user to change the initial parameters of the pendulum swing and the rest of its parameters, resize the canvas, clean, reset, pause, and start the simulation.
7- Visualize simple black thick line representing trajectory on white background, no axes, no text, just trajectory without cleaning.

PROMPT 31. OFF-GRID BACTERIAL CLUSTERS

Write a program in JS/HTML:

1- We work in a continuous space (in a box) of side L.
2- In the center, place a live bacterium (circle) of size R=L/128.
3- From the living bacteria, draw one. Around it, draw a possible space (a free space where there are no other bacteria and where a bacterium at a distance R can be attached). To do this, sample its surroundings by selecting 360/K random angles from 0 to 360 degrees and check if another bacterium can be attached there (if the spot is empty). If there is no space available around the bacterium in question, it dies, and a new bacterium is chosen randomly from the remaining live ones.
4- Attach a bacterium to the place found in point 3.
5- Continue the loop and add more bacteria.
6- Visualize the circuit in two ways
   - the first, white circles with a black border on a white background with no other elements or colors
   - the second, a growth map, showing all growth events as lines that connect the cell around which the next one appeared with the center of this added one – here draw black lines alone on a white background
7- Allow the user to change L, R, K, pause, clean up the layout, and start over.
8- Add a second mode of the simulation: change circular bacteria into elongated cylinders of an x b aspect ratio and modify the algorithm accordingly: seek for space around elongated and randomly rotated shapes, visualize as a rectangle with rounded ends, allow the user to modify a and b.

PROMPT 32. SPRING-MASS SYSTEM (SODA CONSTRUCTOR)

Create a program in a single HTML5/JS file:

1- We will simulate point masses with springs and gravity in 2D.
   1a The damped, Hookean linear spring acts between two connected points.

2- We use Verlet algorithm for integration of equation of motion.
3- Allow the user to add points (masses) using the mouse, add a spring by clicking on a point, dragging the mouse, and releasing the button over another existing point (the rest length of added spring is the length at the time it have been added). During such mouse movement, show the connection.
   - If the user adds a point and then drags the mouse to another place, draw the spring, and when the user releases the button in another place, a second point should be added with a spring connection between those two added points.
5- The user can drag the point by clicking on it.
6- Allow the user to change the type of point to "suspended" by clicking once on it and releasing right after that.
7- Draw in WxH canvas with a floor at H/10 from which points reflect with damping with coefficient f=0.8.
8- Allow user to change (by right click) the type of spring from an ordinary with a fixed equilibrium length to a muscle-type with equilibrium length changing in time as d(t) = d0 + A0 * sin(wt+fi)
9- Allow user to change A0, w, fi for each spring when it is selected and it is a muscle, make edit GUI elements to appear when the spring is selected.
10- Allow user to change W, H, f, Gravity, and integration step dt, spring damping, and spring constant.
11- Allow user to reset, clear, pause, start, enter edit mode (visible as a light gray background), and simulation mode (visible as a white background).
12- Visualization simple, no colors, in 2D, black lines representing springs, dashed if the spring is a muscle, minor points as masses, white background (do not use colors), suspended points in red.

PROMPT 33. SOFT BODY

Create single HTML5/JS file application:

1- Conduct a numerical simulation of the motion of point masses in a gravitational field with springs (Hooke's forces with damping) and with pressure.
   1a- Simulation on a canvas of size WxH with a floor at height H/10 from which material points bounce with damping with coefficient f=0.8.
   1b- The initial condition is N points uniformly distributed on the edge of a circle of radius R. Each point is connected by a spring to its neighbor (a closed loop is formed).
2- Use Verlet's algorithm to integrate the equations of motion of the points.
4- Allow the user to move the points during the simulation (by grabbing with the mouse).
5- At each time step, calculate:
   - gravitational force (for each point)
   - Hooke's elastic force with elasticity k and motion damping constant kd – for each spring connecting two points
   - S – the area of the body (approximated by the rectangle described on it, the area of which will approximate the area of the whole body)
6 In each step, count the pressure as P=C * 1/S, where C is a constant and S is the approximate volume of the body
7 The vector value of the pressure force acting on each point is formed by multiplying P by the vector normal to the point. The vector normal to a point is calculated as the average of the vectors normal to the edges to which that point belongs. The normal vector point outward from the body (as viewed from the center of gravity).

8 Correct the length of all springs. To do this, for each spring with an equilibrium distance d0, calculate the difference between d0 and the current length, and if it is greater than D% d0 (initially D=10) – move the points symmetrically to or from each other so that the distance is equal to D% d0.
9 After calculating the forces in the points 5–7 and integrating the equations of motion with them, and applying the correction to the length of the springs, make a visualization in 2D, black lines representing the springs, minor points as masses, and draw the normal vectors to the points in red.
10 Allow the user to change all the listed parameters using UI: W, H, R, C, f, k, kd, D.
11 Allow pause, restart, etc.

PROMPT 34. BALLISTIC DEPOSITION (CLUSTER)

Write a program in JS/HTML:

1- Work on a WxH grid, with integer values of 0 or 1
2- Place an embryo in the center of the grid
3- Randomly select a point on a circle with center at the center of the layout and radius W
4- Trace the radius from the drawn point to the center of the layout (iteratively, move along the unit vector in that direction multiplied by dx/2, where dx is one pixel). Find the cell of value 1 in which the ray strikes first. Fill in the cell from which the ray came by hitting that cell
5- Return to step 3. Finish if the radius of the resulting cluster reaches W/2.
6- Draw the cluster on a canvas of size bWxbH, black color 1, white color 0
7- Let the user choose W, H, dx, and b. Additionally, optimize your code by taking into account the current cluster radius at the center (e.g., shoot rays from a circle with a radius slightly larger than the current cluster radius, perform adaptive steps, etc.).

PROMPT 35A. TRAFFIC JAM, VERSION 1

(Failed attempt to describe the simulation – see the introduction)
Write a program in JS/HTML:

1- Create a one-dimensional array T[N] (of size N, numbering from 0 to N-1) and integer values greater than or equal to 0. Assume cyclic boundary conditions (consider them for sliding and walking in the following algorithm).
2- Set vmax=5.
3- In an array with probability P=0.5, we insert the value 0 (space on the highway), or put the car by entering its speed V, where V is randomized from the set of integers [1; VMAX].
4- We start the simulation loop:
5- Choose a random number K from the range [0; N-1]. Create a variable A.
6a- If T[K] = 0 (no autos), return to step 4.
6b- If T[K] > 0 (auto with speed T[K]), execute a loop from i=K as long as i <= K+T[K] (with step 1). In each step of the loop, remember A=i. If T[i]!=0 during the walk, break the loop.
7a- If the value of A after executing the loop is K+T[K]+1, it means that we have reached this field without any obstacles. So move the car to a new place: T[A-1] = T[K] and T[K]=0 (in that order). Also, increase its speed, but remember the VMAX limit, i.e., T[A-1]++, and if T[A-1] > VMAX, then T[A-1] = VMAX.
7b- If the value of A after the execution of the loop is less, that is, the condition A < K+T[K]+1 is satisfied – this means that we met an obstacle on the road (most likely another car). If A > T[K]+1 (the case when we are standing in a dense traffic jam and there is just a car in front of us), move the car to a new place, that is: T[A-1] = T[K] and T[K] = 0. In this case, slow down the car, keeping in mind the minimum speed V=1, ie, T[A-1] – and if T[A-1]<1, then

T[A-1]=1. 10-Return to step 5 (continue simulation). 11. Monitor the number of cars on the highway (display for the user). Their variable number indicates a problem with the algorithm.

12- Allow the user to: stop, start, reset (draw), set VMAX, change N, and set P.

13- Visualize the circuit in two ways, black and white, without fireworks and colors. Way 1: as black circles of radius R on a circle of radius r (let the user change r and R) in such a way that where there is a value of 0 in the array, there is no circle, and where there is a non-zero value, a circle is drawn. Draw a large circle with a thin line on the edge. Map the position in the array to an angle on the circle so that N circles fill the full 360-degree angle on the visualization, and the last circle is a neighbor of the first (adjacent angles of 0 and 360 degrees at 12 o'clock). Method 2: as one-dimensional lines of black (non-zero value) and white (zero value in the array) pixels representing the entire array. The lines should be drawn one under the other during the application time at the frequency of line/F time steps (let F = 5 and allow the user to modify F).

PROMPT 35B. TRAFFIC JAM, VERSION 2

Write a program in JS/HTML:

0- Constants: VMIN=1, VMAX=5, DT=1

1- Create an array of size N and integer values >= 0. Apply periodic boundary conditions.

2- Each cell of the array corresponds to a piece of highway. It contains at most one car (it then has a positive value from VMIN to VMAX indicating the car's speed) or a value of 0 (the space is empty and may be occupied by a car).

3- Arrange the random cars in the array randomly with a density of P=0.5. Draw randomly the speed of each parked car.

4- Start the simulation loop:

5- Select randomly one of the cars.

6- If the car has no other cars in front of it on the way, move it by the distance VMAX*DT. Accelerate it (taking into account the VMAX limitation).

7- If it has other cars in front of it, let it reach the nearest car it has in front of it and reduce its speed by 1 (taking into account the VMIN limitation).

8- Perform points 5–7 in a loop.

9- Live list the number of cars.

12- Allow the user to: stop, start, reset (shuffle), set VMIN and VMAX, change N, and set P.

13- Visualize the layout in two ways (simultaneously on two different canvases), black and white, without fireworks and colors. Way 1: as black circles of radius R on a circle of radius r (let the user change r and R) in such a way that where there is a value of 0 in the array, there is no circle, and where there is a non-zero value, a circle is drawn. Draw a large circle with a thin line on the edge. Map the position in the array to an angle on the circle so that N circles fill the full 360-degree angle on the visualization, and the last circle is a neighbor of the first (adjacent angles of 0 and 360 degrees at 12 o'clock). Method 2: as one-dimensional lines of black (non-zero value) and white (zero value in the array) pixels representing the entire array. The lines should be drawn one under the other at the frequency of line/F time steps (let F = 5 and allow the user to modify F).

PROMPT 36. CHROMOSTEREOPSIS

Write a program in JS/HTML:

1- Create a canvas with LxL spacing, black background.

2- In a circle with radius R=L/2 in the center of the layout, randomly draw N1=100 circles with radius r=5px in red color col1=(255, 0, 0). Make sure that drawn points are distributed uniformly across the circle.

3- On the whole canvas, excluding the circle with radius R2=L/2, is placed in the center of the layout, randomly draw N2=100 circles with radius r with blue color col2=(0, 0, 255).
4- Blur the drawn circles gently so that they are not sharp.
5- Allow the user to clean up, draw from scratch, set all the constants and parameters that occur (L, R, R2, N1, r, N2, col1, col2, and the radius of the blurred circles).
6- Let the background of the entire page, UI, and the default canvas color be black.

PROMPT 37. STRANGE ATTRACTOR

Write a single-file program in HTML5:

1. Let $-1<a<1$.
2. We work on a canvas WxH, (800x600).
3. Start at the center of the layout (x0=W/2, y0=H/2).
4. In a loop, perform the transformation N times:
5. x(n+1) = y(n) + h(x(n)), where the numbers in parentheses at the coordinates denote the time step, n is the time step number.
6. y(n+1) = −x(n) + h(x(n+1)).
7. Let the function h(x) be defined as h(x) = a*x+2*(1−a)*x*x/(1+x*x).
8. Visualize straightforwardly by drawing points at S*(x, y)+(xc, yc), where S is the scale (e.g., S=1.0) (draw in groups, e.g., every N2=1000 iterations at a time of N2 points), and xc and yc are the center of the image in screen coordinates. Optimize for drawing as many points as possible without chewing delays. Each point put on the screen should add a C value to the current color on the canvas.
9. Keep the interface simple, all black and white, drawings white on a black background, except for the iterated function and UI; do not draw anything else on the screen.
10. Allow the user to Restart (one button, clears the entire canvas [possibly all tiles in tiled mode] and restarts drawing from the first iteration for the current parameters), also allow the user to set x0, y0, N and change a, set W, H and N2, C.
11. Add "Randomize ON/OFF" button. When pressed, it turns the mode on or off in which the results of the above iterations for different (random) values of a are displayed on the wxh (5x5) grids across the screen in tiles. The visualization also prints out in each tile what a was used there. This button also clears the canvas, sets the iteration number to 1, sets the point in (x0, y0), and starts new iterations for each tile. Calculate the iterations in the tiles one by one (each time setting x0 and y0 to the positions read from the UI). Allow w and h to change.

PROMPT 38. THE LATTICE BOLTZMANN METHOD (LBM)

Write the program in a single JS/HTML5 file:

1- Implement the LBM (the Lattice Boltzmann Method) in D2Q9.
2- Perform a simulation of flow through a channel with a circular obstacle placed in the center.
3- Present the results as a velocity vector field (vectors scaled by a constant), white background, black: vectors, obstacles, particles, and user interface.
4- Use periodic grid in x-axis, and no-slip conditions on top walls, bottom walls, and all obstacles. For no-slip conditions, use the bounce-back technique.
5- Allow change of gravity, accelerating the flow.
6- Allow change of the canvas size and resolution of the LBM mesh in both directions.

7- Use massless particles to visualize the flow trajectory. The particles should initially be arranged randomly and move along the calculated velocity field. If the particle finds itself in a cell with an obstacle or exits the system, insert it at a random place on the grid.
8- Allow to reset the simulation (clear obstacles, velocity field, and distribution function).
9- Allow to add obstacles of the size currently set in the user interface, which appear in a random place when you click "Add". Re-initialize the LBM speed distribution function after adding an obstacle.
10- Calculate on the fly the magnitude of "Tortuosity", i.e. T=<v>/<vx>, where < > means the average across the space available for flow, "vx" is the velocity component in the direction of macroscopic flow (e.g., in the direction of the force acting there), and "v" is the value of velocity at a given point. On the screen, write out how much T. Along with the tortuosity, write out the porosity (the ratio of the area occupied by obstacles to the total area).
11- Allow to modify the number of particles in the system.
12- Allow to modify the relaxation time in the LBM model.
13- Allow modifying the scaling constant of velocity vectors in the visualization (by slider and text field without limits).
14- The code is complicated, so after writing it, check its integrity (all functions, structures, and variables are defined). Correct if something is not right.
15- Be aware of stability of the LBM (Mach number constraint and be sure of proper boundary conditions handling).

PROMPT 39. AGENT-BASED EPIDEMIC SPREADING

Write a program in HTML5:

1- Simulate N agents (circles of radius r) moving in a room of size LxL, with rigid boundary conditions.
2- The circles interact with each other in a spring-like manner (collisions). Use a virtual spring model so that the agents can gently step on each other.
2- Agents move in a uniform rectilinear motion, only changing direction of motion when colliding with other agents.
3- Initially, all agents have a status of "Healthy" represented as a green color in the simulation. N2 agents selected randomly get the status "Sick".
4- When a healthy agent comes into contact with a sick agent, the sickness passes with probability p from the sick agent to the healthy agent.
5- During the simulation, each sick person dies with probability p2.
6- The simulation stops when all agents are sick or have died.
7- During the simulation, draw on a single graph:
Z(t) – healthy since time in green, C(t) – sick since time in red, U(t) – dead since time in black.
8- Allow reset (initial state, clean graphs, set time to 0), pause/start, set L, r, N, N2, p, P2, and other parameters such as collisions.

PROMPT 40. L-SYSTEM TREE

Write a program in HTML5:

1- We are working on a WxH canvas
2- Let state = "-X"
3- Create a map of rules for swapping successive state elements, for example: (X → F+[[X]-X]-F[-FX]+X), (F → FF)

4- Allow a single iteration on the state: the loop goes through all the elements and executes:
   (a) "[" saving the current state (position and angle) on the stack
   (b) "]" removing the state from the stack
   (c) "F" move forward by FWD pixels and draw a line (e.g., FWD=1)
   (d) "+" rotate by ANG degrees to the left
   (e) "-" rotation by ANG degrees to the right
5- Allow cleaning the state, cleaning the canvas, and changing the initial state, rules, ANG, FWD, and canvas dimensions (entered as text strings)
6- The starting point is at the bottom of the canvas, halfway horizontally. The starting direction is "up".
7- Do not implement anything more than what is above, a simple user interface, no colors, etc.

PROMPT 41. DRAGON CURVE

Write a program in single HTML/JS file:

1- We work on WxH canvas.
2- Start with (x, y)=(0, 0).
3- Iterate the following functions (choose f1 or f2 randomly):

f1(x, y) = (1/sqrt(2)) * R(45) * (x, y),

f2(x, y) = (1/sqrt(2)) * R(135) * (x, y) + (1, 0), where R(phi), is a 2x2 matrix of rotation by phi degrees.

4- Draw successive iterations as points on the 2D plane.
5- Allow the user to reset, pause, clean up the image, select other iteration functions (selection of rotation parameters), select the starting point, determine what ranges of (x, y) correspond to the canvas (scaling)
6- Make a simple visualization (black points on a white background), similarly (simple) make a UI.
7- Don't impelement anything more than what is above, simple UI, no colors, etc.
8- Initial ranges: xmin=–0.35, xmax=1.35, ymin=–0.5, ymax=1
9- Optimize for speed.

PROMPT 42. GRANULAR MATTER WITH FRICTION

Write a program in a single HTML5 file:

1- Let d0=1mm, system size 10x5cm.
2- Use implicit Euler and Verlet to integrate the equations of motion.
3- Granules i and j have centers at Ri and Rj, respectively. Let r = Ri – Rj.
4- The vector n is the unit vector parallel to r, namely n = r/|r|, where |r| is the length vector r.
5- The vector s = rot(n, 90), rotation by 90 degrees in the clockwise direction.
6- Velocity v=Vi – Vj, where Vi is the velocity of granule i and Vj is the velocity of granule j.
7- Radius i-th granule is ai.
8- Two granules interact when their distance is smaller than sum of their radii. Force on particle i exerted by particle j in contact is: F(j->i) = Fn * n + Fs * s, where normal comonent Fn is: Fn = kn * (ai + aj – r) ^ (3/2) – γn * me (v * n), kn is the elasticity constant. "^" denotes a power. The first term is the Hertz model. The second term is the velocity-dependent friction term, which is introduced to dissipate energy when bodies collide. γn

is the constant controlling this dissipation, and effective mass me=mi*mj/(mi+mj). The tangential force Fs in the force equation is:

$$Fs = -ɣs*me*(v*s) - sign(ΔS)*min(ks*ΔS, μ*Fn),$$

where the first term is the energy dissipation due to the motion along the s vector. The second term is static friction. When two bodies start contacting we "insert" a virtual spring between them, which extends in the shear direction. For a total displacement ΔS during contact, there is therefore a force ks*ΔS, which acts opposite to the movement of the bodies. The maximum value of this force is given by the Coulomb criterion of static friction μ*Fn. When two bodies are not in contact – the spring is removed as well as the static friction force. For model to work correctly, we must remember the total relative displacement of the bodies in contact in the shear direction and not just the instantaneous displacement between two time steps.

9- Gravity force −mi*g acts on each particle.
10- The total force on a particle is the vector sum of contact with other particles, contact with the wall and gravity.
11- Visualisation: white background, white circles with a 1px wide black border to represent static particles and filled circles in black for dynamic particles, no additional embellishments.
12- Initial values: g=−10m/s^2, μ=0.2, kn=10^6, ks=10^4, ɣn=5*10^2, ɣs=ɣn/100, and for walls kn=2*10^6. Allow user to change physical parameters, size, time step, integrator type, entered numerically in text boxes.
13- Allow to add static (suspended in space but active with collisions and friction on other moving particles) and dynamic particles. Both can be added in two ways. a) Click to start designing a line (starting point), release the mouse to designate the end point. When you release the mouse button, add particles to lay along the line with their edges touching from the start point to the end point. b) Click to start designing a rectangle (start point – top left corner), release the mouse to designate the end point (bottom right corner, or other depending on where you have moved the mouse). After releasing the mouse button, add particles with the same spread as those added in the stream from above in such a way that they fill as precisely as possible (as close as possible to the hexagonal network and dense packing) the designed rectangle touching the edges. In both cases (line and rectangle), when a granule is added, immediately draw the next radius randomly.
14- After pressing the "Circle" button at time t=0, on a circle of radius W/3 (W is the width of the layout), arrange the granules side by side so that they tightly fill the entire circle. These granules move in a rotational motion around the centre of the circle with an angular velocity w. Allow the user to vary w. These rotating granules are not subject to forces, but interact with other granules in the system.
15- Allow the user to select the radius, ang angular speed of the added circle. Particles on the circle that rotate should be drawn as empty circles.
16- All boundaries are open – granules are removed from simulation if they go beyond them.

PROMPT 43. STANDARD MAP

Write a program in HTML5:

1) Start from point (1, 1) and K = 0.971635.
1) Iterate a pair of equations:

$$p(n+1) = p(n) + K * \sin(t(n))$$

t(n+1) = t(n) + p(n+1), where n is the iteration number. Start from the middle. After each iteration, take modulo 2*π of both p(n+1) and t(n+1).

3) Draw p(n+1), t(n+1) on an 800x800 canvas.
4) Allow the user to set K (text box), restart the simulation, clean up and change the canvas dimensions.
5) Allow to change display modes: 1- black dots on a white background, 2- additive drawing of dots (black background, adding the value of A each time the dot is positioned on the canvas), 3- as point 2, but white background and drawing in black (i.e. subtracting A from the colour). In points 2 and 3 allow the user to change A.
6) Allow the user to change the dimensions of the canvas.
7) Add to the program the text boxes N, N2 and the Randomize button which perform N random selections of the starting points p(0), t(0) and N2 following iterations drawing new points into the layout. Draw live each time the plot is updated.

PROMPT 44. EXCITATION MODEL AND SPIRAL WAVES

Write a program in HTML5:

1- LxL square grid.
2- At each node of the grid, place a tree with probability p1.
3- Use the cellular automaton technique. Work on two copies of the array. Perform the following rules first for all cells in array A writing the result in array B and then copy the whole array B to array A.
4- In a loop:
   - On each cell without a tree and without a wall and not in a refractive state, a tree can grow with probability p2
   - Any cell with a tree can ignite from an edge cell where there is a fire with probability depending on the direction of the ignited neighbour (neighbourhood of 8 directions – 4 main directions and a diagonal)
   - Each tree can ignite itself with probability p4.
   - The tree burns for P simulation steps
   - After burning, the field is in a state of refraction for REF steps in a state where nothing can grow there and only then does it turn into an empty field where something can grow.
5- Quick real-time simulation, simple visualisation: empty field brown, field with tree green, field with fire red, field of wall black, no grid, grid grid colours alone.
6- Allow the user to light trees with the mouse or to draw and remove walls (as lines). Add option "remove all obstacles).
7- Allow to set p1, p2, ignition probabilities from a given direction, p4, P, L (in the range of 50–300) – by slider or typing in values, and resetting the simulation. The simulation should react live to changes made.
8- Allow the boundary conditions to change to periodic and back to boundary fields with walls (can't burn, can't grow trees there).

PROMPT 45. PAPER, STONE, SCISSORS

Write a program in a single HTML5 file:

1- Simulate a cellular automaton with two WxH arrays labelled as (A, B), where we work on array A, execute an algorithm on all elements writing the results to B and then make a copy of B->A.

2- For each node, write one of three random values: type1=1, type2=2 or type3=3.
3- In the simulation, consider that type1 beats type2, type2 beats type3 and type3 beats type1.
3- Simulation loop:
4- From the environment of the current node (periodic conditions), with type TA, count how many cells of other types (TB and TC) are in total, the numbers are NA and NB:
   4a) If TB beats TA and NB > THR -> set the new value in the cell to TB
   4b) If TC beats TA and NC > THR -> set new value in cell to TC
   If none of the conditions are met: leave the current value in the cell.
5- Return to 4.
6- Allow start/pause, change W, H, change THR (initially 2)
7- Optimise the simulation code (speed of operation).
8- Use simple rendering, no grid, just grids representing colours on canvas, use red (255, 0, 0), green (0, 255, 0) and blue (0, 0, 255) for colouring.

PROMPT 46. CLASSICAL BILLIARDS

Write the program in a single JS/HTML5 file:

1- Work on a 1280x720 canvas.
2- On the canvas, draw a large figure with a thick line: oval, ellipse, circle, square, rectangle, square with a full circle in the center (circle with radius rk excluded from the square such that particles may move only outside of the circle and inside of the square) or rectangle with rounded corners with radius of curvature r1.
3- At the point (x, y), place N points with a random variation of +/– (dx, dy).
4- Give all points the same initial velocity equal to a vector with coordinates (vx, vy).
5- The points move in a uniform rectilinear motion.
6- The points bounce off the edge of the figure drawn in the center of the system.
7- Moving points leave a trace (do not automatically clean).
8- Allow user (GUI, text boxes) to change dx, dy, N, x, y, vx, and vy, time step dt for integration of equations of motion, radius of circle rk, radius of curvature r1.
9- Allow pause, reset, start, and cleanup.
10- Allow selection of shapes and their sizes to be drawn in the center from a drop-down list.
11- Use simple visualization and allow the selection of a color palette for points (gradients, all black, all blue or transparency with adjustable alpha).
12- Allow changing all values in any range (use text fields).
13- If the user selects N=2, add a graph showing the distance of the points in time and a second one on which the Lyapunov exponent for this pair of points is calculated on the fly (drawing with live theory fitting).
14- For collisions, take into account precisely the shape of the figure – for example, for a rectangle with a radius of curvature, take into account the rounded corners so that the points reflect precisely from these rounded corners. Make sure all points remain within the shape (collision of particles being inside of the shape with shape edge should reflect them back inside of the shape).
15- Use a white background in the visualization.

PROMPT 47. PLANETARY SYSTEMS

Write a program in a single JS/HTML file:

1- Work on an 800x800 canvas.
2- Place a large mass M in the center of the system.

3- Arrange eight smaller masses with masses m1.m8<<M arranged farther and farther (equally spaced) from the central mass. Give them a small velocity tangent to a hypothetical orbit (e.g., perpendicular to the segment connecting the mass to the center of the system).
4- Integrate the equations of motion for all masses in the system using the Verlet method.
5- Assume that there is a law of universal gravity between bodies i and j, and the resulting force acting on both bodies has a different sign: Fij = −Gmimj * (1/(rij^2+eps))*rij/|rij|, where rij=ri-rj is the vector from body j to body i, eps is a small number (softening factor). Add the constant epsilon ("softening factor") to eliminate the problem of body collisions.
6- Visualize simply – black small circles on a white background.
7- Allow user to pause, start, reset, and set m1 to m8, M, epsilon, integration time step dt, and start positions of all masses.
8- Allow to turn drawing of trajectories on/off and set their maximum lengths (as a text field – unlimited), also allow to clear them.
9- Allow adding and removing masses and create different configurations.

PROMPT 48. DIFFUSIVE TORTUOSITY

Write the program in a single JS/HTML file:

1- Set N wanderers in the center of the layout r(0)=(x0, y0) on square canvas 500x500.
  1.1 In the layout, draw NB square-shaped obstacles of linear size A (over the entire available simulation area).
2- In a loop after time t:
3- In a loop after 50 wanderers i:
  3.1 Take a step to one of the nearest neighbors ri(t)=ri(t–1)+dr, where dr is a random offset in one of the four directions. If there is an obstacle at the target location, do not make the move.
  3.2 Draw the trajectory of the point ri(t).
  3.3 As you move, calculate on the fly the mean square distance of the simulation from the starting point MSD(t) = <|r(t) – r(0)|^2>=(1/N)*Sum(| ri(t)-r(0) |^2), where the sum runs over all wanderers indexed by i.
4- Draw the trajectories of all the wanderers.
5- On a separate canvas, draw separately the graph of MSD(t) (for the entire simulation time) and numerical fit of the Smoluchowski-Einstein theory MSD(t)=2dDt to the simulation data of MSD(t) by numerical fitting of the function f(t)=2dDt, by least squares method, to MSD(t) data obtained from simulation. Assume d is the dimension, t is the time, and D is the diffusion coefficient. On the graph, label the time and MSD values with numbers.
6- Simple visualization, on a white background, wanderers in black, but allow the user to choose a palette for coloring the trajectory (different colors, gradient, one selected color – choose red, blue and black, alpha and set its level), allow changing x0, y0, N, NB, A, reset, pause, start, setting alpha for obstacles. Use simplistic GUI style, no colors, etc.
7- Optimize the simulation and let you set its speed.

PROMPT 49. GRAY-SCOTT DIFFUSION REACTION

Write the programm in a single JS/HTML file:

1- We work on WxH computational grids (initially 800x600).
2- On the grids we hold (scalar fields) A, Anew, B, and Bnew.

3- The initial values of A and B are: A=1, B=0 (in the whole area) with a slight disturbance in the central part of 20x20, where A=0.5, B=0.25.
4- After initialization, add a random value from −dr to dr.
5- In a loop, for each cell, perform the operation: Anew = A + dt * (DA * ΔA – AB^2 + F*(1 – A)), Bnew = B + dt * (DB * ΔB + AB^2 – (F + k)*B).
6- Use the 5-point formula to expand the Laplace operator Δ, e.g. ΔA ≈ A(x+1, y) + A(x–1, y) + A(x, y+1) + A(x, y–1) – 4A(x, y))/h^2.
7- Visualize on the canvas. Show fields A and B using field values to multiply separate palettes added together. Let you choose the color palette for A and B from an extensive range and possibilities (from trivial single color palettes, gradients to complex palettes and generated from several trigonometric functions for an interesting set of colors).
8- Add to the visualization the possibility of post-processing, calculating normal, and counting spotlight (bump-mapping technique), and let you choose the parameters of this post-processing.
9- Optimize heavily the simulation speed to be dynamic and real-time.
10- Allow the user to change W, H, dr, DA, DB, F, k, and dt (through sliders but also by typing values into text boxes without limiting their values).
11- Let the user set a gradient value for the F and k parameters such that their values change linearly from min to max along a selected axis (vertically or horizontally) from one edge of the layout to the other. Allow the user to choose the mode of selection of these two variables (either constants or gradients)
12- Use a minimalist GUI without colors.

PROMPT 50. ANT COLONY

Write a program in a JS/HTML file:

1- In the lower left corner of the canvas, WxH=200x150, place a circle with a radius of W/8; this will be the anthill. Randomly place N=50 ants in the anthill area.
2- Ants move in a fixed direction (each has a velocity vector with a modulus of 1) and have a floating-point position (despite showing the ant on the grid, mathematically this object can have non-integer positions). Use periodic boundary conditions or reflection conditions (as a parameter).
3- The ant periodically performs a smooth, random rotation of the direction vector. Allow the user to specify the rate (0.1), range (45 degrees), and weight W(0.1) of these changes.
4- The ant leaves a trail behind it (hereinafter referred to as pheromone).
Store it in a pheromone accumulation array (floating point type) by adding the EPS value (parameter) at the location where the ant is currently located. There will be two types of ants: those going for food (type A) and those returning to the anthill (type B). Hence, there will be two types of pheromones: FA and FB, which are left behind by the respective types of ants. Store information about the pheromone of a given type of ant separately.
5- Pheromones decay over time (exponentially or linearly, as desired) and are subject to diffusion (with a parameter – diffusion constant).
6- In the upper right corner, on the grid, mark a circle with a radius of R=W/8. This is the food area.
7- The ant's field of vision is limited to the angle attached to its head, which collects information in front of the ant (allow the user to change the angle and range of perception).
8- If an ant in state A (walking towards food) has food within its field of vision, it smoothly turns towards it with weight W2 (1.0). When a food field is directly below its head, it captures it, and from that moment on, it enters food-carrying mode, continuing to move while

carrying the food particle on its head. This particle also disappears from the food field. After capturing the food, the ant begins a walk in which it smoothly turns 180 degrees and then, with a new direction, continues its journey to the anthill.

9- Design the cave. To do this, set all cells that do not belong to a rectangle with dimensions (W-W/32, H-H/32) with rounded corners with a radius of curvature SIGMA (dimensions and angle to be set by the user) to wall cells. No ant can enter the cave area (or any other obstacle).
10- If an ant senses a cave wall at a distance of WALLDISTANCE (allow the user to change this value), rotate its direction of movement at a speed that ensures smooth rotation but is also fast enough to avoid the obstacle or move in a different direction (the parameter is the rotation speed).
11- If an ant carrying food enters the anthill, it leaves the food there (the food disappears). The anthill counts and displays (text) the amount of food it has accepted. After leaving the food in the anthill area, a new ant appears in a random free space in the anthill. After leaving the food, the ant begins a walk in which it rotates 180 degrees, and then continues its journey for food in the new direction.
12- The ant adjusts its angle of movement to follow one type of pheromone. Its direction vector changes towards the highest concentration of pheromones of the opposite type (ant A is attracted to FB pheromones, and ant B is attracted to FA pheromones).
13- Ants avoid each other and do not overlap, but this should be done smoothly, without a rigid condition (use, for example, virtual springs that gently repel the ants from each other if they get too close, with the spring constant set by the user).
14- If an ant in state B (returning to the anthill) has the anthill in its field of vision, it smoothly turns in its direction with weight W3.
15- Add the ability to switch to edit mode, which pauses the simulation and allows you to draw with the mouse, using brushes of various thicknesses: obstacles for ants, anthills, place ants, and add food.

Additional Information

Visualization: light brown background in the simulation, dark brown obstacles, orange anthill, light green food, pheromones as a so-called heat map displayed throughout the simulation with a color palette based on the color of food for pheromones left by ants carrying food and with a color palette based on the color of the anthill for ants searching for food (mix the colors of both pheromones). Ants are small, 6x2 resolution objects with six single-pixel legs and simple animation during movement (alternating flashing). Allow changing the colors of pheromones and heatmap properties in the GUI.

GUI: Allow the user to start, pause, clear the pheromone board, and scale the visualization (make sure the canvas does not cover the GUI). The user can change the parameters: ant speed, pheromone decay rate, and diffusion constant, as well as the size of the pheromone trail left by the ants, the width and height of the board where the pheromone data is stored, and the rate of change of the ant animation. Make sure that the GUI area does not overlap with the simulation area. Add a Reset button that resets the fields and resets the ants (but does not change the settings selected in the GUI). Allow the user to change all numerical parameters within a wide range by placing both sliders and text fields in the GUI (allowing unlimited number entry).

PROMPT 51. WAVE EQUATION

Write a program in a single JS/HTML file:

1- Work on a WxH grid.
2- Place a point in the center of the array that will smoothly oscillate, according to the sin(t) function.

3- The rest of the grid will value will be computed from equations below.
4- Solve the wave equation numerically in time, according to Eq.:

∂^2 u(r, t)/∂t^2=-v^2*(∂^2u/∂^2x^2 + d^2u/∂y^2),

where u(x, t) is the deflection field, the sign "^" denotes the derivative, and v denotes the speed of propagation of the wave.
5- For solving the above equation use the method of splitting the wave equation into two equations by introducing a new variable, namely:

E=∂u/∂t

which gives simpler form of wave equation for the deflection speed E:

∂E/∂t = −v^2*∂^2U/∂t^2

Discretize the above two equations for E on a numerical grid. Use nine-point analog to discretize the Laplace operator. Discretize time derivatives using a 1st-order difference. From E, compute U (deflection) and draw the result. The above can be summarized as:

E(n+1, i) = E(n, i)+v^2 * lapl(U) dt

U(n+1, i) = U(n, i)+E(n+1, i)*dt,

with lapl(U) – the differential analog of the Laplace operator. Do not use damping in the formulas.
6- Use damped boundary conditions on the edges (choose several methods and give the user a choice, let MUR and PML be among these methods).
7- GUI: Allow resetting, clearing obstacles, starting over, changing mesh parameters (W, H), rescaling, and visualization and wave parameters, and setting new obstacles by drawing sections with the mouse and setting (adding and removing) wave sources with different parameters. Allow numerical values to be entered using both sliders and text boxes (no numerical restrictions). Make sure that both canvas and GUI area are well visible and accessible to the user.
8- Visualization: visualize the layout as a deflection (height) color map (a two-dimensional drawing in which the color of a pixel corresponds to the deflection U). Use interesting palettes (user selectable). Also allow switching to 3D mode, in which the waveform is represented by: a) a white grid on a black background, b) as in a), but including the removal of lines covered by surfaces (do not show lines belonging to backward-facing surfaces even though these are theoretically transparent), c) visualization with shadow and lighting, d) visualization with a simple generated texture.

Verify JS code and validate it to be able to execute it in the browser. Use math.pow instead of **. Don't use shaders, just plain JS.

PROMPT 52. ELECTRON IN A MAGNETIC FIELD

Write a program in a single JS/HTML5 file to run in a browser:

1- We are working in an 800x600 layout.
2- An electron (q=−1) starts from the point x0, y0 with an initial velocity V0=(V0X, V0Y, V0Z).

3- There is a constant magnetic field B=(BX, BY, BZ) in the system, initially B=(0, 0, 1).
4- Let us write down the equation of motion of the electron:

d^2 r/dt^2 = F/m, where

F is the force acting on the electron, m is its mass, and the second derivative on the left-hand side represents the acceleration.
5- Integrate the equation of motion of the electron, taking into account the Lorentz force (interaction of the moving electron with the magnetic field), i.e., F=q*(v x B), where q is the electron's charge, v its velocity, and B the magnetic field. Use the Velocity Verlet, constant-energy method with additional acceleration computed in each time step on top of temporary updated velocity values (make sure to not introduce make numerical errors in this procedure).
6- Visualize the motion of the electron and the path of motion.
7- Visualize in 2D, white background, taking into account the position of Z as the size of the point (and the width of the trajectory).
8- Allow N electrons to start with a random scatter of initial DR position or DV velocity (to be set by the user) and draw trajectories of all.
9- Minimalist, simple GUI, no colours in the user interface, with the ability to manually enter parameters into text boxes (in addition to sliders), and start, pause, and reset. Allow users to change DT, x0, y0, V0X/Y/Z, and visualization parameters (colors, path length, transparency, and thickness of trajectory and point – all using gui elements and text fields). Allow separate B components to be set as a linear gradient from minimum to maximum across the width, height, or depth of the layout separately for each field component. Allow selection of the number of N points.
10- Allow to set NK (initially 1) – the number of integration steps per frame of simulation displayed (in order to be able to speed up calculations without the need to increase DT).
11- Use initial parameters: dt=0.01, N=10, x0=70, y0=50, z0=0, DR=0, V0X=60, V0Y=20, V0Z=0, DV=4, BXmin=0, BXmax=1 (gradient in x), BYmin=0, BYmax=0.25 (gradient in y), BZmin=0, BZmax=1 (gradient in z), path length = 21200.

PROMPT 53. SMOOTHED PARTICLE HYDRODYNAMICS (SPH)

Write a program in a single JS/HTML file to run in a browser:

1- Work in a 2x1.5 (WxH) layout, where (0, 0) lies in the lower left corner.
2- Distribute evenly NX to NY points in rectangle (0, 0) – (W/A1, H/B1) with additional random spread dx, dy.
3- Define the interpolation kernels and their gradients (the "^" sign is a power-up): W_poly6(d) = 315/(64πh^9)(h^2 – d^2)^3 for d<=h and W_poly6(d) = 0 for d>h, W_poly6_grad(d)= 945/(32πh^9)(h^2 – dd)^2, W_poly6_grad_x(rx, ry, d) = –rx * W_poly6_grad(d), W_poly6_grad_y(rx, ry, d) – –ry * W_poly6_grad(d),

W_poly6_lapl(d)= 945/(8πh^9) * (h^2 – d^2) * (d^2 –3/4(h^2–d^2)),
W_spiky_grad(d)= 45/(πh^6d) (h-d)(h-d),

W_spiky_grad_x(rx, ry, d) = –rx * W_spiky_grad(d),

W_spiky_grad_y(rx, ry, d) = –ry * W_spiky_grad(d),

W_viscosity_lapl(d)= 45/(π*h^5) * (1 – d/h).

In kernels, ensure secure values of d = 0 or d close to 0 to prevent kernel values from exploding.

4- For each particle i, calculate its density by calculating the sum over its j neighbors for which r<=h. To do this, calculate the expression ρi = sum(j) m*W_poly6(r, h), where r= |ri – rj| (also consider the contribution of the ith molecule itself).

5- For each i-th molecule, clean the force acting on the molecule and the gradient and Laplacian accumulator, and then make a loop over its j neighbors (for which r<=h) to add the forces from that molecule and calculate the surface tension properties:

   5.1 Constrain ρ and ρn to a maximum value of ρ0, i.e., ρ=max(ρ, ρ0) and ρn=max(ρn, ρ0). Calculate the pressures: p = k*(ρ – ρ0), pn = k *(ρn – ρ0), where ρ and ρn are the density of the particle and its neighbor.

   5.2 Add the pressure force Fp = –(m*(p+pn)/(2*ρn)) * W_spiky_grad(r) (use separate versions for the x and y components of the gradient)

   5.3 Add viscous force Fvisc = (ηm(vn – v)/(ρn)) * W_viscosity_lapl(r), where v and vn are the velocities of the particle and its neighbor, respectively

   5.4 Add gradient: (m/ρn) * W_poly6_grad(r) to the gradient accumulator (use separate versions for the x and y components of the gradient)

   5.5 Add laplasian: (m/ρn) * W_poly6_lapl(r) to the laplasian accumulator.

6- Total force F = Fp + Fvisc.

   6.1 Acceleration a = (F/ρ) + g

   6.2 Velocity v = v + a * dt

   6.3 Position x = x + v * dt

7- Default values of the variables h = 0.02, ρ0 = 10, η = 10.0, gy = –1.0, dt = 0.0025, k = 8.0, particle mass m=0.003, random dx=0.00001, random dy=0.00001.

8- At walls and obstacles, reflect particles at a distance of DS=4*h (soft particle radius) and consider this distance when generating a column of points so that they are not generated in the collision area.

9- Implement an additional velocity smoothing model in which, after counting the forces, a correction is made for each i-th particle before updating the position: vi = vi + epssum(j) (mj/rhoij)(vj-vi)W_poly6(r, h), where rhoij=0.5(ρi+ρj), and sum after j runs on the neighbors of molecule i (for which r<=h). 10- Allow the user to change all simulation parameters from the GUI using sliders and text boxes to set numeric variables in wide, unlimited ranges. Allow changing the height and width B1 and A1 of the generated column of points, as well as the number of NX and NY particles. Allow changing the scaling of S and make sure the canvas and GUI area are clearly visible. Keep the GUI minimalist, without colors or fireworks. Allow speed smoothing correction to be enabled and disabled. Allow to change DS.

11- Visualize black particles on a white background. Scale the layout by a factor of S for visualization (default S=400). Allow changes in visualization properties, visualization of points, their colors, and a selectable marching cubes (squares in 2D) algorithm for finding fluid surfaces (allow setting its properties and make sure that the algorithm results in smooth surfaces whose quality depends on the grid resolution and threshold of values selected by the user). Make sure both visualizations (surface and points) are visible and can be combined together or shown separately. Additionally, enhance the visualization near the edges to prevent the marching cubes algorithm from generating erroneous lines. Draw everything in 2D. When drawing obstacles, take into account the DS modification and draw to show the physical edges of objects.

12- Add a design mode where the user can add particles with a set density, add no-slip obstacles, clean them up, and reset the scene to a basic state.

PROMPT 54. HYDRAULIC TORTUOSITY (RESEARCH)

Write an application as a single JS HTML file:

1- Implement the Lattice Boltzmann method for fluid flow simulation in a channel with periodic boundary conditions in x and y directions. Use BGK and TRT relaxations in the D2Q9 version. Allow the user to select one of them and change their properties (i.e. change relaxation time in BGK, etc.).

2- Accelerate the fluid by gravity in the horizontal direction x.
3- Place obstacles at random position (the position of obstacles center point in the range 0, 0 to NX, NY). Obstacles have a no-slip condition (velocity=0) at their boundaries. Obstacles may have different shapes and similar sizes in the channel. Keep adding them until the porosity is 0.7 (calculated for the whole domain). Add another deposition algorithm, in which the obstacles do not overlap and have a minimum edge-edge distance A. Obstacles may extend outside of the system and be partially beyond the edge, in that case they should be drawn periodically on the other side (both in x and in y direction)
4- Visualize the flow with 4000 massless particles following the velocity field. Remove and place those particles that penetrate or bump into obstacles. Place them in a new, random location.
5- When calculating, calculate the tortuosity from the velocity field as the mean velocity modulus divided by the mean velocity in the x-direction. Also, calculate the Reynolds number and permeability from Darcy's law.
6- Add sliders so that the user can vary the external force from minimum to maximum in any range (with the possibility of entering its value from a text box).
7- At each time step, recalculate steady state condition by calculating the maximum change in tortuosity between a two time steps at times t and t+50. Display it as a single number like the other data.
8- Add a button to randomly place obstacles and a slider to change the overall porosity of the configurations generated with this button. Add the possibility to change parameter A – minimum distance of two obstacles from each other. Also, add the possibility to change the type and radius of the obstacles. In the procedure for drawing new positions, include the mechanism that prevents the algorithm from clogging. There is a limit on the porosity that cannot be lowered further. Thus, introduce a condition: if NM times you try to place an obstacle, but it cannot be done, you end up and accept a higher porosity.
9- Add a slider to control the delta time used for the time integration of massless particles in the animation (with the possibility to type manually in the text box, value unlimited).
10- Set the size of the computational grid to NXxNY (300x200). Allow NX and NY to vary. Keep the width W of the visualization canvas as 800 and adjust the height to the aspect ratio of the calculation grid. Allow to set S – the degree of scaling of the canvas.
11- Add a canvas next to the simulation with a graph of the tortuosity measured over time. This graph should reset every time the user presses the random configuration button.
12- Make sure the simulation is stable, test the CFL condition.
13- Add a text box next to the workspace with 20 lines.
14- Add a "save" button that adds data to the text box in the format: simulation step, porosity, tortuosity, permeability (from Darcy's law), convergence criterion, Reynolds number – measured from the current velocity field.
15- Add the possibility to perform a series of measurements in which we set the range and step of variation of porosity phimin, phimax, and phistep. Allow user to set the number of simulation steps to be performed for each simulation. Allow user to set the number NPOR repetitions for one porosity. Once you have clicked start, run cycle of simulations. Each run should use a different random seed, and save the results to the text box you have created. Limit the number of attempts to embed obstacles for a given porosity. If NM times you cannot embed a new obstacle, leave the porosity as it is. Create a second "Averaged results" text box to display an averaged results for a given porosity. Hence, each result in this second text box is the average of the NPOR values calculated earlier in the simulation process. Average porosity as well. In addition to the average, immediately following it by a space, also write the calculated standard error equal to the standard deviation divided by the root of the NPOR number. Thus, number of data in one line of the second text field will be doubled.

16- Add a clean field that clears the text box.
17- Visualize points, velocity field as color map (grayscale) with the brightest color representing the largest velocity modulus. Draw obstacles in dark turquoise.
18- Place GUI on the left side and ensure that the canvas area (on the right) does not overlap with it. Allow the user to start, pause and reset the simulation (a complete new start with all initializations).

PROMPT 55. SHALLOW WATER

Write the program in a single JS/HTML file:

1- We are working in one dimension in a system of width (number of nodes) n.
2- Let u(x, t) be the velocity of the fluid column at point i at time t.
3- Let h(x, t) be the height of the fluid column.
4- Let b(x, 0) be the height of the substrate (Earth's surface) (constant over time).
5- Determine the initial moment h(x, 0), h(x, 1) and the shape of the substrate b(x, 0).
6- Let h(x, 0) be a straight line of H/2 with a pulse of amplitude equal to the height of the system/10.
7- Let h(x, 1) be a straight line of H/2 with an impulse of amplitude equal to the height of the system/10.1.
8- Let b(x, 0) be a straight line with a value of H/12.
9- Start the calculation loop after times (start with j=2 with incrementation by 1):
10- If there are any sources or sinks of water in position i, add or subtract water from the current value of h(i, j).
11- Calculate for all columns and values d(i, j) = h(i, j–1) – b(i, 0), if d(i, j)<0, then d(i, j)=0.
12- Calculate the new value of h(x, j) from h(x, j–1) and h(x, j–2) from Eq: **A***h(i, t)=h(i, t–1)+(1–tau)*(h(i, t–1)-h(i, t–2)), where tau is a value from 0 to 1 and represents the viscosity of the fluid. Matrix **A** is a three-diagonal matrix, where on the diagonal lie the elements e0, e1, e2, . . ., en–1. In the diagonals lying above and below the central diagonal, the elements f0, f1, . . ., fn-2 are placed. The elements are calculated from the formulas:

e0 = 1 + g * (dt*dt) * (d(0, t) + d(1, t))/(2*dx*dx), and

ei = 1 + g * (dt*dt) * (d(i–1, t) + 2*d(i, t) + d(i+1, t))/(2*dx*dx), fora(0 < i < n–1), and

en–1 = 1 + g * (dt*dt) * (d(n–2, t) + d(n–1, t))/(2*dx*dx), and

fi = –g*(dt*dt)*(d(i, t) + d(i+1, t))/(2*dx*dx).

13- After calculating the new values of h(x, t), correct new h(x, t) (only those spots that are not 0) to keep the volume of the fluid constant between two steps.
14- If for any index i the value of h(i, t) < b(i, 0) set hi(j) = h(i, j–1) = b(i, 0)-eps, where eps is a small epsilon.
15- Simple visualization, no graphic effects, no colors in the GUI. Visualize the solution on a WxH (800x600) canvas by drawing the black ground from 0 to b(x, 0) and the water column from b(x, 0)+h(x, t) in dark turquoise. From b(x, 0)+h(x, t) to H draw in light blue.
16- Allow the user to set all possible simulation parameters, change colors, grid resolution, canvas resolution, start, stop, reset, and design initial values, initial fluid pulses, and bottoms in a wide range and in a flexible way, giving the possibility to design interesting fluid layouts. Also, allow adding and removing fluid sources and sinks in the layout.

PROMPT 56. MOLECULAR DYNAMICS

Write the simulation in a single JS/HTML file:

1- Arrange N classical atoms on a two-dimensional surface by arranging them in a triangular lattice structure with distances arranged in a minimum potential. Let the network of atoms macroscopically form a structure as close to a circle as possible.

2- The interaction potential between two atoms:

$$V(r)=4\varepsilon[(\sigma/r)^{12} - (\sigma/r)^{6}]$$

3- Use reduced units: $\sigma=\varepsilon=m=kb=1$.

4- The size of the rectangle in which the simulation takes place is 20x20. Use rigid boundary conditions (the atom cannot penetrate the walls of the system).

5- Use Verlet's velocity method to integrate the equations of motion of atoms. For each i-th atom, calculate:

$$\text{New positions: } ri = ri + vi\,dt + (1/2)ai(dt\,dt),$$

$$\text{New velocities: } vi = vi + (1/2)ai\,dt,$$

New accelerations ai, where the force acting on a particle is calculated from the Lennard-Jones potential gradient (equation above) for all particles adjacent to it within a selected radius rc=3 (the so-called cut-off radius),

$$\text{New velocities: } vi = vi + (1/2) * ai * dt.$$

6- ri – position vector of the i-th atom, vi – velocity of the i-th atom, dt – integration time step,
7- During the simulation, calculate the total energy of the system from the Lennard-Jones potential and from the velocity (sum of kinetic and potential energy)

8- Draw the initial values of the velocities of the atoms from the Maxwell-Boltzmann distribution for a fixed temperature, which reduces to a normal distribution for the velocity components. e.g., for the x-component, for a fixed temperature T, it will be:

$$f(vx) = \mathrm{sqrt}(m/(2\pi kbT))\exp(-mvxvx/(2kbT)).$$

Use the Box-Muller method or an analogous method to draw numbers from a Gaussian distribution.

9- After drawing the velocity, make sure that the total momentum of the system is zero (calculate the velocity of the center of mass and subtract it from any velocity in the system).

10- Monitor the temperature of the system on an ongoing basis by averaging and scaling the kinetic energy of the atoms accordingly. Show the instantaneous temperature as a graph over time.

11- Conduct a simulation in constant energy (E), volume of the system (V), and number of molecules (N).

12- Visualize atoms as points with a visualization radius rw.

13- Add a mode to visualize the color of points depending on their velocity with mapping to several color palettes to choose from (palettes to choose from a wide range of styles).

14- Add an initial state design mode where the user can: restart the state, remove selected atoms with a brush of a certain radius, add atoms in the brush area in places so that they do not overlap with other atoms, add atoms so that they are docked to neighboring atoms in the brush area so that they are in the minima of the potential (a triangular network should be formed in this mode).

15- During the simulation, allow the user to increase and decrease the volume of the box, change the temperature of the system (velocity scaling), change the sign of the time step dt with a single button, and add a gravitational force acting downward in the y-axis.
16- Also give the user the ability to edit all the constants in the simulation.
17- Simple, minimalist GUI without embellishments, visualization on a 1280x720 canvas with scaling.

PROMPT 57. PREDATOR PREY SYSTEMS

Write a program in a single HTML5 file (to be opened in a browser):

1- Solve numerically the system of differential equations: dx/dt = ax – bxy, dy/dt = –cy + exy, where a, b, c, and e are the parameters of the model.
2- Use the Euler method to solve, where after discretizing the derivatives in time: x(n)=x(n–1) + (ax(n–1) – bx(n–1)y(n–1)) * dt, y(n) = y(n–1) + (–cy(n–1) + e*x(n–1)*y(n–1)) * dt,
3- Use the following initial conditions and parameters: x(0)=y(0)=10, a=1.1, b=0.4, c=0.1, e=0.4, dt=0.01.
4- Draw on one graph dynamically the course of x(t) together with y(t), always showing the whole graph from time t=0.
5- Next to it on the second graph, draw the function x(y).
6- Allow reset, pause, start, clean up graph 1, clean up graph 2, and change all model parameters.
7- Clean the x(y) graph only when the special "clean x(y)" button is pressed, because that is where the results are supposed to be drawn.
8- Use a minimalist GUI interface, without colors. Similarly, graphs should be black and white (with a white background).

PROMPT 58. HEAT TRANSFER

Write the program in a single JS/HTML file:

1- Work in a layout with NX resolution on NY (300x200).
2- Create grids: T with temperature, TWK specifying wall conductivity.
3- Start with TWK = 1 in the whole area except for a central frame of 2 grids wide by NX/1.5 side length with hollow holes of 4 grids wide at the corners. For this frame, set conductivity = 0.1.
4- Begin by initializing T with a Gaussian distribution in the center of the layout.
5- Iterate the equation in time:
T(i, j, t+1)=T(i, j, t)+ TWK(i, j) * (T(i–1, j, t) + T(i+1, j, t) + T(i, j+1, t) + T(i, j–1, t))/h^2) * Δt.
In iteration, use the technique of two copies of T grids so that the values are not overwritten. Only after passing the entire grid using the above equation, make a copy to the target array.
6- Visualize the temperature distribution over time using a color map (mapping and color palette to be selected by the user from several available palettes) and/or a constant temperature contour map with levels automatically selected by the program (the user then provides the number of levels) or by user selection of levels. Allow mixing of visualizations: mode 1) black-on-white contours alone with obstacles drawn, a legend of temperature values written into the contour with text slanted along the contour, mode 2) like the first, but in addition a color map, mode 3) colors alone, mode 4) arrows showing the local temperature gradient in the layout (both black and white and combinations with color map and temperature), mode 5) all modes at once.

7- When drawing contours and arrows together with the color map, draw the color map first, and then draw the arrows and contours. When designing schematically (gray level), mark the conductivity map. On the final visualization, draw these values with the corresponding permeability level.
8- Allow the user to scale the visualization canvas, change simulation parameters (including NX and NY), add walls, remove them, specify two parameters: temperature and conductivity TWK, and temperature for the wall being drawn, start, pause, reset, clean T, set initial T with a Gaussian with specified parameters, and speed up the simulation. Let the program allow simple drawing of walls, either with segments or as a free brush drawing with specified thickness.
9- Let the numerical data be changed with sliders and by typing values into text boxes.
10- On visualization with contours, let you set the number of text tags describing the contour.
11- When dividing (normalizations, etc.), make sure that you do not divide by 0.
12- When writing JS code, analyze the correctness of the program, syntax, and completeness so as not to generate errors.

PROMPT 59. CROWD EVACUATION

Write the program in a single JS/HTML file (to be opened in a browser):

1- Work on an 800x400 board representing an 80x40m room.
2- In the middle of the room, on a vertical diagonal, place a wall with a door opening 2m wide in the middle of the wall.
3- Draw the positions of N people on the left side of the room so that they do not overlap. Each ith person occupies the area of a circle of radius Ri=1m.
4- The equation of motion describes the motion of each person in the arrangement:

$$m \cdot dvi/dt = m \cdot (v0 \cdot ei0(t) - vi(t))/\tau i + sum(j, j \neq i)\ fij + sum(W)\ fiW,$$

where vi is the motion vector of the i-th person, v0 is the maximum walking speed, ei0(t) is the unit vector indicating the direction of the i-th person's movement (to the emergency exit), and τi is the characteristic time of matching the person's vi(t) speed to the target speed. The forces of interaction between the i-th and j-th person are described by the formula:

$$fij = (Ai \cdot exp((Rij-dij)/Bi) + k \cdot g(Rij-dij)) \cdot nij + \kappa \cdot g(Rij-dij)/(\Delta vtji \cdot tij),$$

where Ai, Bi are constants, Rij is the sum of the radii of the two persons, dij is the difference between the positions of the two persons dij=|ri – rj|, nij is the normalized vector from person j to person i, that is, the vector nij = (nij1, nij2)=(ri – rj)/dij, tij is a unit vector perpendicular to nij, that is, tij=(-nij2, nij1), the difference of the perpendicular components of velocity Δvtji is Δvtji=(vj-vi)·tij, and k and κ are constants. g(x) is a function, taking the value g(x)=0 for x<0 and g(x)=x for x>0. The last term of the forces, fiW, we determine for the interaction with the wall. For the distance diW from the wall and the vector niW indicating the direction to the wall and tiW (direction perpendicular to niW), we have the formula:

$$fiW = (Ai \cdot exp((ri - diW)/Bi) + k \cdot g(ri-diW)) \cdot niW - \kappa g \cdot ((ri-diW)/(vi \cdot tiW)) \cdot tiW$$

5- Use initial values of constants: mi=80, v0=1, τi=0.5, Ai=*210^3, Bi=0.08, k=1.*210^5, κ=2.4*10^5.

6- When the user presses “Alarm”, forces are activated to propel all people toward the exit (door).
7- After entering the door area, the person moves further attracted by the center of the opposite wall in the other room.
8- Use a simple visualization: walls as lines, doors as wider rectangles, people as circles with the radius of each person’s area, with the direction of movement from the center of the circle to the surface marked (line), use only black edges on a white background.
9- Allow the user to change all the constants and parameters of the simulation, physical constants, maximum gait speed, reset the state, move, add and subtract people individually or in larger groups (add them randomly in free spaces), draw walls and add, remove, move and change the width of the door, accelerate, decelerate, stop and restart the simulation (accelerate as repeated execution of a computational loop, keep a fixed time step selected by the user). Allow the user to change the size of the canvas for visualization, the physical dimensions of the room, and the radius of the area occupied by one person.
10- The GUI should allow you to enter numbers both with a slider and from a text box. Make sure all elements are large enough to allow free input.
11- Create a new graph canvas and draw a real-time plot of the number of people in the room against time over the entire range (from zero to the current time in the simulation. Label the axes as time (x) and number of people (y), and dynamically mark the maximum and minimum quantities on the axes. Allow resetting (cleaning) of this graph.

PROMPT 60. SIMPLE (COMPUTATIONAL FLUID DYNAMICS)

Write the program in a single JS/HTML5 file:

0- Let us define the following equations: Equation 1: ∂u/∂t = – ∂u^2/∂x – ∂uv/∂^y – ∂p/∂x + (1/Re)·(∂^2u/∂x^2+∂^2v/∂y^2), Equation 2: ∂v/∂t = – ∂v^2/∂y – ∂uv/∂^x – ∂p/∂y + (1/Re)·(∂^2u/∂x^2+∂^2v/∂y^2), Equation 3: ∇^2P’ = (1/dt)(∇·V).
1- Create, and clear arrays of P, P*, U, U*, and V, V*, representing pressure, and velocity components. Keep all in two copies for two time steps.
2- Use staggered grid of NX by NY resolution with velocity components U placed at the top and bottom edge, V on the left and right edge and pressure P in the center of each node.
3- Use the following boundary conditions: left edge, bottom edge, and right edge are all no-slip. Top lid, from i=1 to i=NX-2 (excluding corners that remain no-slip) have constant velocity condition U=u0, V=0.
4- Solve equations (1) and (2) for the fields U*(n+1) and V*(n+1). Use discrete numerical analogous for differential operators on staggered grid to achieve second oreder accuracy. Use forward explicit discretization for the time derivative, second-order central discretization for second derivatives, forward discretization for pressure derivatives.
5- Solve the Poisson equation (3) to get P’. As with velocities, use a differential grid for discretization and compute the solution using the Gauss-Seidel iterative method with successive overrelaxation (SOR).
6- Calculate U’ and V’.
7- Compute pressure in the next time step: P(n+1) = P*(n) + P’.
8- Compute velocity in the next time step U(n+1)=U*(n)+U’ and V(n+1)=V*(n)+V’, where U’ and V’ values are calculated from simplified momentum equations: U’=-Δt·(∂P’/∂x) V’=-Δt·(∂P’/∂y).
9- Visualize the velocity field, velocity vectors or N massless particles advected with a DTP time step. Massless particles are advected using velocity from bilinear interpolation of velocity field. In the bilinear interpolation algorithm, take into account staggered grid arrangement of data in the grid (keep 0 at boundaries and u0 at top moving-lid). Do not

take values from outside of the domain if you get close to the boundary. Particles that stuck at boundary should be placed in randomly in the system. Allow turning on and off the cleaning such that particle may be smeared on the canvas.

10- User may select different visualization modes or combine them. Particles are visualized as white points on a black background with floating positions at a specified alpha level (transparency, particle size chosen by the user). The smallest particle size should be 1px in the canvas resolution. Add an option to reflect the canvas in the x-axis and y-axis. Reflection should affect both particle and field visualization.

11- Allow the user to change dt, dx, NX, NY, u0, N, DTP, and visualization properties (palette for velocity, particle style), and the scaling size of the canvas for visualization (when changing properties do not reset the simulation but continue from the current velocity). Allow the user to turn off clearing the canvas, start, pause and reset. Allow numbers to be entered via text boxes (without limits) as well as with sliders.

12- User may specify Reynolds number (which is used directly in the momentum equation) by changing viscosity, u0, dx, dy, NX (Re=u0L/viscosity, where u0 is top-lid velocity, L=NXdx, and viscosity is set by the user).

13- Initially use: dt=0.001, dx=dy=0.01, u0=0.5, NX=100, NY=100, kinematic viscosity=0.01.

PROMPT 61. SPHERE TRACING

Write the program in a single JS/HTML file:

1- Work on a 1280x720 resolution canvas.

2- In a loop, for each pixel (px, py), iterate the ray marching algorithm:

2.1 Determine the vector p0 from the eye located at (0, 0, 0) to the pixel on the screen at position (px/RX, py/RY, Z0), where RX, RY is the resolution of the canvas, and Z0 is the position of the viewport in the z-axis.

2.2 Let the vector p initially be p = p0 (the starting point).

2.3 Let the vector dir be equal to p0. Normalize dir (dir=dir/|dir|).

2.4 Execute an iterative loop (maximum ITER times):

2.4.1 Calculate the distance d=f(p).

2.4.2 If d<ε, break the iterative loop.

2.4.3 Move by distance d, i.e., p=p+dir*d.

2.5 If the loop has executed ITER times, the color of the pixel is black. Go back to step 2 (continue with pixels).

2.6 If the loop executed less than ITER times, calculate IL = scalar product between the normal vector calculated from the gradient of the distance function at point p, and the vector indicating the position of the light located at point (lx, ly, lz). Count the light intensity: light=(1.0+0.5*IL)/distlight^2, where distlight is the distance of point p from the light source.

2.7 Perform the new iteration of ray marching (steps 2.2–2.4.3) from point p to the light source (lx, ly, lz) and count the sum of the smallest distances from other objects in the scene when going from p to (lx, ly, lz). This sum will be our shadow value, which, when trimmed to 0–1, multiplies our light. The final expression for light is thus:

lightfinal =light * shadow, and this will be the color (brightness) value of the pixel.

3- Perform the above algorithm for the distance function f(x), which returns the minimum distance between the objects of the sphere at the point (0, 0, –10) and the surface with orientation (0, 0, 1). Alternatively, add a second mode for merging objects that uses the smoothstep function instead of the minimum, allowing one object to merge and transition smoothly into another.

4- The camera is set so that the direction (0, 0, 1) points upward, and the camera faces a sphere located at pnct (0, 0, −10), that is, in the visualization, a surface with such a
   The normal vector will be horizontally aligned. With this camera configuration and scene, we should see a ball in the center, located on the surface.
5- Draw the scene described earlier using the above algorithm.
6- Add the ability to rotate the scene (camera) around the x, y, and z axes with the mouse.
7- Add the ability to change the parameters of background lighting and light source in the scene.
8- Add the ability to add and remove cube and sphere objects anywhere in the scene.
9- When changing parameters, render only an approximate image at a lower resolution to keep the application interactive. Only when you click "Render" should you draw a high-resolution image.
10- Add simple post-processing (brightness, contrast).
11- Add the ability to change the parameters of the ray tracing procedure.
12- Allow the user to choose how to merge objects: min-distance function or smoothstep for a smooth transition (let the user choose the type of merging and its parameters).
13- Allow the parameters in the GUI to be changed with sliders and to type values into text boxes.

PROMPT 62. STALAGMITE GROWTH

Write a program in a single JS/HTML file:

1- We are working in array A with dimensions NX by NY.
2- Initially, clear the array (0), then fill the entire row from 1 to NX at height y=0 with the value 2 (substrate).
3- Set the position of the particle to px=NX/2, py=NY.
4- In the time loop, perform:
   4.1 Movement with probabilities: p1 for movement py=py−1, p2 for horizontal movement (equally possible either px=px+1or px=px−1). If, after drawing a movement, the molecule would land on a field with A different from 0, reverse the movement.
   4.2 Deposition, if there is a substrate under the molecule (value 2), i.e., A(px, py−1)=2, then with probability p3 set A(px, py)=2, where the value of p3 is calculated using the formula:

   $$p3 = a + b*(n\cdot g)\text{^}c,$$

   (n is a vector normal to the surface, pointing outside the structure, the '*' sign is multiplication, the '·' sign is scalar product, the '^' sign is exponentiation). The vector g=(0,−1), and 'a', 'b', and 'c' are constants. Calculate the normal vector to the surface as the normalized sum of the normal vectors to the unit faces in the approximate staircase located in a circle with a radius of NC grids from (px, py). After depositing the substrate, place the molecule in its initial position.
   4.3 If the molecule did not settle in the previous step, but there is a substrate underneath it, it performs a horizontal movement (sliding) in the direction of the lowest column. To do this, it checks its surroundings within a radius of NCSLIZG to find which side has the lowest-height column, and moves in that direction by one field with probability p4. If the levels on the sides are the same but lower than the height of the column in which the molecule is currently located, it performs a random movement to the right or left.
5- Visualize array A on a canvas scaled by S (pixel magnification). Colors: 0 in the array is black, 2 in the array is white. Also, draw a suitably scaled particle at position (px, py) in light turquoise.

6- End the simulation when the particle is deposited on the surface (setting the value to 2) in the row with position y=NY–3.
7- Allow start, pause, reset, manual addition of subsequent particles, and single-step execution.
8- Allow changing parameters: NX, NY, NC, S, p1, p2, a, b, and c, NCSLIZG (slider and by typing values into text field, without restrictions on range, to be entered in GUI text fields). Responsive GUI (simulation reacts to entered values in real time).
9- Allow setting the number of steps per display cycle without animation delays (fast simulation mode) – also by specifying the number in text field.
10- Allow turning on/off and scaling the visualization of calculated normal vectors.
11- Allow you to turn on/off visualization in which you color the structure by mapping the time of attachment to a color from the palette (provide at least 5 different palettes for the user to choose, including from blue to white, arctic style).

PROMPT 63. JETPACK – A COMPUTER GAME WITH PHYSICS

Write the program in a single JS/HTML file (to be opened in a browser):

1- Work on a WxH canvas (e.g., 1280x720).
2- The game can have two modes: START, PLAY, END.
3- In START mode, write that the game titled "Jetpack" will start when you press the spacebar. After pressing it, the game goes into PLAY mode.
4- The next part refers to the PLAY mode. At the height of H/20 from the bottom surface of the layout (bottom in the sense of displayed at the bottom of the canvas as the user sees it) is a horizontal surface.
5- Place an astronaut on the surface. The astronaut's height is H/20, and he has a space backpack with jet engines on his back. The two jet engines are at the bottom of the backpack and point with their outlets toward his legs. There are also side engines on the backpack. The astronaut is described by position r and velocity v.
4- Next to the astronaut is a rocket (empty), which is 2.5x the height of the astronaut.
5- The user controls the astronaut. The down arrow button activates the jet engines at the bottom of the backpack, which eject mass in the negative y-direction in the astronaut's reference frame, causing the astronaut to begin floating. The left arrow turns on the motor facing left in the astronaut's reference system, and the right arrow turns on the motor facing right. Activating the motor causes the mass to be ejected in the right direction and the astronaut to move in the opposite direction. The motors on the sides only work if the astronaut is in the air.
5- Add randomly over the surface, distributed small islands. On these islets, fuel appears randomly in time (with WP frequency). If the hero enters the fuel area, it attaches to their backpack (visible in the visualization).
6- The astronaut can drop the fuel at any time by pressing the spacebar. After pressing the spacebar, the fuel slowly begins its descent phase to the surface of the ground or an islet, doing so from the drop point in a smooth, downward direction until it stops. If the fuel touches the rocket, the fuel level in the rocket increases (from an initial value of 0), which can be seen in the visualization as the successive coloring of the rocket with the orange color of the fuel from below. Then, the hero also receives a maximum of 100 points for the fuel brought, but the number of points decreases linearly to 1 over time, starting the countdown from the moment the fuel is collected.
7- If the fuel fills the entire rocket and the astronaut stands on the ground in the rocket area, an animation of departure into space is performed. The astronaut flies off into space. The animation should last a minimum of 10 seconds, counting down the points and smoothly

transitioning to the end of the game and the win message. Then, the user can enter their name and be added to the list of top scores.

8- Visualize the game as a vector graphic by drawing a slightly undulating mountainous terrain as a white line on a black background. The character should also be vectorized with precise details. When buttons are pressed and jet engines are started, let particles emerge from the engines to visualize the ejection of matter. The matter ejected from the engine is also vectorized, with objects fading in and out of existence over time until they expire.
9- Add an approximation effect. When the gas button (up arrow) is pressed, the camera zooms very slowly towards the character, and when the button is released and momentum is lost, the camera zooms away. The movement should be smooth (slow-fast-slow). Keep an eye on specific minimum and maximum values, of course, so that the camera does not get too close or too far away.
10- Optimize the simulation and particle system so that the simulation and animation are smooth and do not slow down.
11- Add randomly placed asteroids with a specific TA frequency. Like fuel, they are located on the surface side of the astronaut (above the surface). If the hero enters them, he loses 1/10 of his energy.
12- Display at the top of the screen (in the bar) the passing time, the amount of fuel collected, and the energy bar.
13- When all energy is lost, the hero dies. Display summary information and let you start again by pressing the spacebar.
14- When the game ends for any reason, the game goes into END mode. Give the player the option to write his name and record a list of the top 10 scores. Remember the list of records the next time you start the game. Equivalent to the decision to enter a name and save the score, the player should be able to simply start the game again by pressing the space bar.
15- When visualizing, the astronaut is always positioned at the center of the canvas (both horizontally and vertically), and the environment moves (using a technique where the game camera is always positioned over the astronaut, so that when he moves, the camera moves with him).
16- Choose speeds, physical constants, and jet engine properties so that the animation is smooth and the user can control the astronaut freely. The game should not be challenging. Use low gravity and non-sensitive engines. The player should be able to fly away freely.
17- Allow the user (GUI displayed next to the game canvas) to set all the parameters that change the properties of the game – speeds, constants, frequencies of fuel and asteroid appearances, power to change angles and rotations, engine powers, etc. Numerical magnitudes allow you to change manually or using text boxes.
18- Pay attention to the state of the game (the welcome screen, the game itself, the final screen) and adjust what is displayed to that state. Take care to initialize objects at the right time to avoid creating problems when running JS code.

PROMPT 64. CHAIN FOUNTAIN (MOULD EFFECT)

Create a two-dimensional simulation in a single JS/HTML file:

1- Create a model of a chain in a gravitational field.
2- The chain is made of beads connected by Hooke springs with high elasticity and some damping depending on the speed difference between the beads.
3- Do not use additional constraints on the distance between points.
4- Make sure that the chain is always completely arranged in the container at the initial moment, regardless of how large the container is and where it is located. The initial distances between the beads should be balanced.

5- The user can grab any point and move it. Clicking a point adds a virtual spring with a high elasticity constant that pulls the point toward the mouse position.
6- Design a container without a top wall. All other walls of the container are made of circles arranged closely to create the illusion of almost straight walls, and the tops of the side columns are rounded.
7- Implement a bead-wall collision model where points penetrating any of the circles forming the walls are repelled with a spring force proportional to the depth of penetration, with a constant k2 (a virtual spring appearing upon collision).
8- Allow the user to set the number of beads N, their radius R, distance from each other D, mass M, gravity g, distance of the container from the surface below it onto which the chain can fall, height of the container, elasticity constant k of the connections on the chain and k2 for collisions, radius (width) of the vessel walls. The user can enter variables using text fields (without min/max restrictions).
9- Use the Verlet method to integrate the equations of motion of the points. Allow the time step in the method to be set.
10- Use simple visualization: black beads and segments, white background, black container walls. Show the endpoints of the chain in a different color.
11- If the mouse is over one of the ends, it catches that chain as a priority.
12- Allow the container to be positioned (moved) on the x and y axes.

Bibliography

Auer, S. *Realtime Particle-Based Fluid Simulation*. Technische Universität München, 2009.

Bak, P., Chen, K., & Creutz, M. "Self-Organized Criticality in the 'Game of Life'." *Nature* 342.6251 (1989): 780–782.

Baram, R. M., & Herrmann, H. J. "Random Bearings and Their Stability." *Physical Review Letters* 95.22 (2005): 224303.

Baumann, R. Grafika Komputerowa. Wydawnictwa Komunikacji i Łączności, Warszawa 1989.

Benettin, G., Gallavotti, G., Jona-Lasinio, G., & Stella, A. L. "On the Onsager-Yang-Value of the Spontaneous Magnetization." *Communications in Mathematical Physics* 30 (1973): 45–54.

Biggins, J. S. "Growth and Shape of a Chain Fountain." *Europhysics Letters* 106.4 (2014): 44001.

Biggins, J. S., & Warner, M. "Understanding the Chain Fountain." *Proceedings of the Royal Society A: Mathematical, Physical and Engineering Sciences* 470.2163 (2014): 20130689.

Brandt, S., & Dahmen, H. D. Mechanika kwantowa w obrazach. PWN, 1989.

Bridson, R., Houriham, J., & Nordenstam, M. "Curl-Noise for Procedural Fluid Flow." *ACM Transactions on Graphics (ToG)* 26.3 (2007): 46-es.

Caps, H., Dorbolo, S., Ponte, S., Croisier, H., & Vandewalle, N. "Rolling and Slipping Motion of Euler's Disk." *Physical Review E* 69.5 (2004): 056610.

Chen, J. X., & da Vitoria Lobo, N. "Toward Interactive-Rate Simulation of Fluids with Moving Obstacles Using Navier-Stokes Equations." *Graphical Models and Image Processing* 57.2 (1995): 107–116.

Chenciner, A., & Montgomery, R. "A Remarkable Periodic Solution of the Three-Body Problem in the Case of Equal Masses." *Annals of Mathematics* (2000): 881–901.

Chopard, B., & Droz, M. *Cellular Automata Modeling of Physical Systems*. Cambridge University Press, 1998.

Christoffel, K. M., & Brumer, P. "Quantum and Classical Dynamics in the Stadium Billiard." *Physical Review A* 33.2 (1986): 1309.

Cohen, Y., Devauchelle, O., Seybold, H. F., Yi, R. S., Szymczak, P., & Rothman, D. H. "Path Selection in the Growth of Rivers." *Proceedings of the National Academy of Sciences* 112.46 (2015): 14132–14137.

Collett, T., Graham, P., & Heinze, S. "The Neuroethology of Ant Navigation." *Current Biology* 35.3 (2025): R110–R124.

D'Alessio, S. "An Analytical, Numerical and Experimental Study of the Double Pendulum." *European Journal of Physics* 44.1 (2022): 015002.

Dasgupta, S., Stauffer, D., & Dohm, V. "Boundary Effects in the Three-Dimensional Ising Model." *Physica A: Statistical Mechanics and Its Applications* 213.3 (1995): 368–375.

Duda, A., Koza, Z., & Matyka, M. "Hydraulic Tortuosity in Arbitrary Porous Media Flow." *Physical Review E* 84 (2011): 036319.

Eden, M. "A Two-Dimensional Growth Process." *Proceedings of the 4th Berkeley Symposium on Mathematical Statistics and Probability* 4 (1961).

Farkas, I., Helbing, D., & Vicsek, T. "Mexican Waves in an Excitable Medium." *Nature* 419.6903 (2002): 131–132.

Gibney, E. "Physicists Explain 'Gravity-Defying' Chain Trick." *Nature* 14523 (2014).

Gleick, J. (2018). Chaos: narodziny nowej nauki. Poland: Zysk i S-ka Wydawnictwo.

Gray, P., & Scott, S. K. "Autocatalytic Reactions in the Isothermal, Continuous Stirred Tank Reactor: Isolas and Other Forms of Multistability." *Chemical Engineering Science* 38.1 (1983): 29–43.

Harlow, F. H., & Shannon, J. P. "The Splash of a Liquid Drop." *Journal of Applied Physics* 38.10 (1967): 3855–3866.

Hart, J. C. "Sphere Tracing: A Geometric Method for the Antialiased Ray Tracing of Implicit Surfaces." *The Visual Computer* 12.10 (1996): 527–545.

Hecker, C. "Physics, the Next Frontier." *Game Developer* (1996): 12-20.

Heermann, Dieter W. *Podstawy symulacji komputerowych w fizyce*. Wydawnictwa Naukowo-Techniczne, 1997.

Helbing, D., Farkas, I. J., Molnar, P., & Vicsek, T. "Simulation of Pedestrian Crowds in Normal and Evacuation Situations." *Pedestrian and Evacuation Dynamics* 21.2 (2002): 21–58.

Helbing, D., Farkas, I. J., & Vicsek, T. "Simulating Dynamical Features of Escape Panic." *Nature* 407 (2000): 487–490.

https://cargocollective.com/sagejenson/physarum (opened: 2025-11-02)

https://en.wikipedia.org/wiki/Ant_colony_optimization_algorithms (opened 2025-10-03)
https://en.wikipedia.org/wiki/Compartmental_models_(epidemiology) (opened: 15-06-2025)
https://en.wikipedia.org/wiki/Curve-shortening_flow (opened 2025-11-02)
https://en.wikipedia.org/wiki/Discrete_Laplace_operator (opened 2025-10-19)
https://en.wikipedia.org/wiki/Dynamical_billiards (opened 2025-09-20)
https://en.wikipedia.org/wiki/Heat_equation (opened 2025-10-19)
https://en.wikipedia.org/wiki/Karplus-Strong_string_synthesis (opened 2025-11-02)
https://en.wikipedia.org/wiki/Lotka%E2%80%93Volterra_equations (opened 2025-10-18)
https://en.wikipedia.org/wiki/L-system (opened 2025-06-15)
https://en.wikipedia.org/wiki/Conway's_Game_of_Life (opened 2026-03-27)
https://en.wikipedia.org/wiki/Millennium_Prize_Problems (opened 2025-06-28)
https://en.wikipedia.org/wiki/Nagel – Schreckenberg_model (opened 2025-06-07)
https://en.wikipedia.org/wiki/Sierpiński_triangle (opened 2025-05-01)
https://en.wikipedia.org/wiki/Standard_map (opened 2025-06-28)
https://github.com/mattpalermo/chain-fountain-simulator/ (opened 2025-12-01)
https://iquilezles.org/articles/raymarchingdf/ (opened 2025-10-27)
https://paulbourke.net/fractals/randomtile/ (opened 2025-05-05)
https://pl.wikipedia.org/wiki/Krzywa_Lissajous (opened 2025-05-07)
https://www.cs.ubc.ca/~van/sssjava/javademo.html (opened 2025-11-02)
https://www.dynamicmath.xyz/fourier-epicycles/ (opened: 2025-11-02)
https://www.grc.nasa.gov/www/k-12/airplane/socforce.html (opened 2025-11-02)
https://www.reddit.com/r/blender/comments/13oro3p/ive_added_a_bunch_of_new_features_to_my_ant/ (opened 2025-10-02)
https://www.shadertoy.com/ (opened 2025-10-27)
https://www.youtube.com/watch?v=_dQJBBklpQQ (opened 2025-12-01)

Ihmsen, M., Orthmann, J., Solenthaler, B., Kolb, A., & Teschner, M. "SPH Fluids in Computer Graphics." In Eurographics 2014 – State of the Art Reports, The Eurographics Association (2014): 21–42.

Karplus, K., & Strong, A. "Digital Synthesis of Plucked-String and Drum Timbres." *Computer Music Journal* 7.2 (1983): 43–55.

Kass, M., & Miller, G. "Rapid, Stable Fluid Dynamics for Computer Graphics." *Proceedings of the 17th Annual Conference on Computer Graphics and Interactive Techniques*. 1990.

Kim, Y., Russell, J., Karpinska, M., & Iyyer, M. "One Ruler to Measure Them All: Benchmarking Multilingual Long-Context Language Models." *arXiv preprint arXiv:2503.01996* (2025).

Koza, Z., Matyka, M., & Khalili, A. "Finite-Size Anisotropy in Statistically Uniform Porous Media." *Physical Review E* 79 (2009): 066306.

Krüger, T., Kusumaatmaja, H., Kuzmin, A., Shardt, O., Silva, G., & Viggen. "The Lattice Boltzmann Method." *Springer International Publishing* 10.978–3 (2017): 4–15.

Kucharski, I. Symulacja ruchu samolotu w jednorodnym polu grawitacyjnym z wizualizacją 3D. Praca Licencjacka, 2019.

Lander, J. "Research on the Rhine: Reflections on Water Simulation." *Game Developer* (2000): 1–4.

Lander, J., & GRAPHIC CONTENT. "Oh My God, I Inverted Kine." *Game Developer Magazine* 9 (1998): 9–14.

Langton, C. G. (1986). "Studying Artificial Life with Cellular Automata." *Physica D: Nonlinear Phenomena* 22 (1–3): 120–149.

Lee, J., & Herrmann, H. J. "Angle of Repose and Angle of Marginal Stability: Molecular Dynamics of Granular Particles." *Journal of Physics A: Mathematical and General* 26.2 (1993): 373.

Matuszak, M. *Zastosowanie modelu Helbinga w symulacji ewakuacji. praca* magisterska (UWr), 2010.

Matyka, M. "Numerical Modeling of Fluid Flow through Porous Media (Modelowanie numeryczne transportu plynow przez osrodki porowate)." *arXiv preprint arXiv:0907.3132* (2009).

Matyka, M. "Sediment-Water Interface Flow with the Multiparticle Collision Dynamics." *Transport in Porous Media* 120 (2017): 633–641.

Matyka, M. "Solution to Two-Dimensional Incompressible Navier-Stokes Equations with Simple, Simpler and Vorticity-Stream Function Approaches. Driven-lid Cavity Problem: Solution and Visualization." *arXiv preprint physics/0407002* (2004).

Matyka, M. Symulacje Komputerowe w Fizyce, wydanie 2, Helion, Gliwice, 2021.

Matyka, M., Khalili, A., & Koza, Z. "Tortuosity-Porosity Relation in the Porous Media Flow." *Physical Review E* 78 (2008): 026306.

Matyka, M., & Ollila, M. "Pressure Model of Soft Body Simulation." Proceedings of SIGRAD (2003): 325–333.

May, R. M. "Simple Mathematical Models with Very Complicated Dynamics." *Nature* 261.5560 (1976): 459–467.

Mendoza, M., Succi, S., & Herrmann, H. J. "Flow through Randomly Curved Manifolds." *Scientific Reports* 3.1 (2013): 3106.

Monaghan, J. J. "Smoothed Particle Hydrodynamics." *Reports on Progress in Physics* 68.8 (2005): 1703.

Müller, M., Heidelberger, B., Hennix, M., & Ratcliff, J. "Position Based Dynamics." *Journal of Visual Communication and Image Representation* 18.2 (2007, April 1): 109–118.

Nagel, K., & Schreckenberg, M. "A Cellular Automaton Model for Freeway Traffic." (PDF). *Journal de Physique I.* 2.12 (1992): 2221.

Patankar, S. *Numerical Heat Transfer and Fluid Flow*. CRC Press, 2018.

Pearson, J. E. "Complex Patterns in a Simple System." *Science* 261.5118 (1993): 189–192.

Potter, D. E. Metody obliczeniowe fizyki: fizyka komputerowa. Państwowe Wydawnictwo Naukowe, 1982.

Provot, X. "Deformation Constraints in a Mass-Spring Model to Describe Rigid Cloth Behaviour." In *Graphics Interface*. Canadian Information Processing Society, 1995.

Rafler, S. "Generalization of Conway's 'Game of Life' to a Continuous Domain-Smooth Life." *arXiv preprint arXiv:1111.1567* (2011).

Reynolds, C. W. "Flocks, Herds and Schools: A Distributed Behavioral Model." *Proceedings of the 14th Annual Conference on Computer Graphics and Interactive Techniques*. 1987.

Ristow, G. H., & Herrmann, H. J. "Density Patterns in Two Dimensional Hoppers." *Physical Review E* 50.1 (1994): R5.

Schroeder, D. V. "Interactive Molecular Dynamics." *American Journal of Physics* 83.3 (2015): 210–218.

Seybold, H., Andrade Jr, J. S., & Herrmann, H. J. "Modeling River Delta Formation." *Proceedings of the National Academy of Sciences* 104.43 (2007): 16804–16809.

Stam, J. "Real-Time Fluid Dynamics for Games." *Proceedings of the Game Developer Conference* 18.11 (2003).

Stevens, H. "Why Outbreaks Like Coronavirus Spread Exponentially, and How to "Flatten the Curve." *Washington Post*, 2020. https://www.washingtonpost.com/graphics/2020/world/corona-simulator/ (opened 2025-06-15)

Succi, S. *The Lattice Boltzmann Equation: For Fluid Dynamics and Beyond*. Oxford University Press, 2001.

Sznajd-Weron, K., & Sznajd, J. "Opinion Evolution in Closed Community." *International Journal of Modern Physics C* 11.06 (2000): 1157–1165.

Szymczak, P., Ladd, A. J., Lipar, M., & Pekarovič, D. "Shapes of Ideal Stalagmites." *Proceedings of the National Academy of Sciences* 122.42 (2025): e2513263122.

Tessendorf, J. "Simulating Ocean Water." Simulating Nature: Realistic and Interactive Techniques. *SIGGRAPH* 1.2 (2001): 5.

Tokarsky, G. W. "Polygonal Rooms Not Illuminable from Every Point." *The American Mathematical Monthly* 102.10 (1995): 867–879.

van de Panne, M., & Lee, C. "Ski Stunt Simulator: Experiments with Interactive Dynamics." *Proceedings of the 14th Western Computer Graphics Symposium*, Banff, Canada, April 2003.

Wang, C. Y., Liu, P. L., & Bassingthwaighte, J. B. "Off-Lattice Eden-C Cluster Growth Model." *Journal of Physics A: Mathematical and General* 28.8 (1995): 2141.

Wejchert, J., & Haumann, D. "Animation Aerodynamics." *ACM SIGGRAPH Computer Graphics* 25.4 (1991): 19–22.

Welch, J. E., Harlow, F. H., Shannon, J. P., & Daly, B. J. "The MAC Method-a Computing Technique for Solving Viscous, Incompressible, Transient Fluid-Flow Problems Involving Free Surfaces." *Los Alamos Scientific Lab, University of California, New Mexico* (1965): No. LA-3425.

Witkin, A., & Kass, M. "Spacetime Constraints." *ACM Siggraph Computer Graphics* 22.4 (1988): 159–168.

Witten Jr, T. A., & Sander, L.M. "Diffusion-Limited Aggregation, a Kinetic Critical Phenomenon." *Physical Review Letters* 47.19 (1981): 1400.

Wolfram, S., & Gad-el-Hak, M. "A New Kind of Science." *Applied Mechanics Reviews* 56.2 (2003): B18–B19.

Xu, H., Huang, H., & Chen, Y. "Comparison of Absorbing Boundary Conditions of FDTD in High Frequency Electromagnetic Field Simulation." *2009 9th International Conference on Electronic Measurement & Instruments*. IEEE, 2009.

Yu, B., Cui, F., Pei, S., Yingle, Z., & Yifei, L. "Modelling Analysis and Application on Bead Chain Fountains." *European Journal of Physics* 47.1 (2026): 015004.

Index

L

M

N

O

P

Q

R

S

T

U

V

W